HURON COUNTY LIBRARY

Date Due

APR 28 1998			
MAY 27 1999			

BRODART Cat. No. 23 233 Printed in U.S.A.

DISCARD

5218

595
.789
Gra

Grace, Eric, 1948-
The nature of monarch butterflies : beauty takes flight / Eric S. Grace. --Vancouver : Greystone Books, c1997.
114 p. : col. ill., col. map. --(Greystone nature series)

Includes Internet addresses (p. 110), bibliographical references (p. 109-110) and index.
846130 ISBN:1550545701

(SEE NEXT CARD)

253 97OCT20 3559/he 1-489213

THE NATURE OF

MONARCH BUTTERFLIES

THE NATURE OF

MONARCH BUTTERFLIES

Beauty Takes Flight

ERIC S. GRACE

GREYSTONE BOOKS
Douglas & McIntyre
Vancouver/Toronto

OCT 28 '97

Text copyright © 1997 by Eric S. Grace
Photographs copyright © 1997 by the photographers credited

97 98 99 00 01 5 4 3 2 1

All rights reserved. No part of this book may be reproduced, stored in a retrieval system or transmitted, in any form or by any means, without the prior written permission of the publisher or, in the case of photocopying or other reprographic copying, a licence from CANCOPY (Canadian Reprography Collective), Toronto, Ontario.

Greystone Books
A division of Douglas & McIntyre Ltd.
1615 Venables Street
Vancouver, British Columbia V5L 2H1

Originated by Greystone Books and published simultaneously in the United States of America by Sierra Club Books, San Francisco.

CANADIAN CATALOGUING IN PUBLICATION DATA

Grace, Eric, 1948–
The nature of monarch butterflies

(The nature series)
Includes bibliographical references and index.
ISBN 1-55054-570-1

1. Monarch butterfly. I. Title. II. Series: Nature series (Vancouver, B.C.)
QL561.D3G72 1997 595.78'9 C96-910801-X

The quotation on pages 2–3 is from *The Monarch Butterfly* by Fred Urquhart (Toronto: University of Toronto Press, 1960). Reprinted by permission of the publisher.

Jacket and book design by DesignGeist
Editing by Anne Norman
Front jacket photograph by Gavriel Jecan / Art Wolfe, Inc.
Back jacket photograph by Michael Evan Sewell
Printed and bound in Hong Kong by C&C Offset Printing Co., Ltd.

Pages ii–iii: Photograph by Frans Lanting / First Light
Page iv: Photograph by Wayne Lynch
Page viii: Photograph by Robert McCaw

The publisher gratefully acknowledges the assistance of the Canada Council and of the British Columbia Ministry of Tourism, Small Business and Culture.

To Oliver,
who initiated
my metamorphosis
from a scientist
to a science writer

CONTENTS

	ACKNOWLEDGEMENTS	xi
	INTRODUCTION	1
PART 1	**THE LIFE OF A MONARCH**	7
PART 2	**MIGRATION**	39
PART 3	**THE MONARCHS' WORLD**	67
PART 4	**THE NEED FOR CONSERVATION**	87
	FOR FURTHER READING	109
	INDEX	111

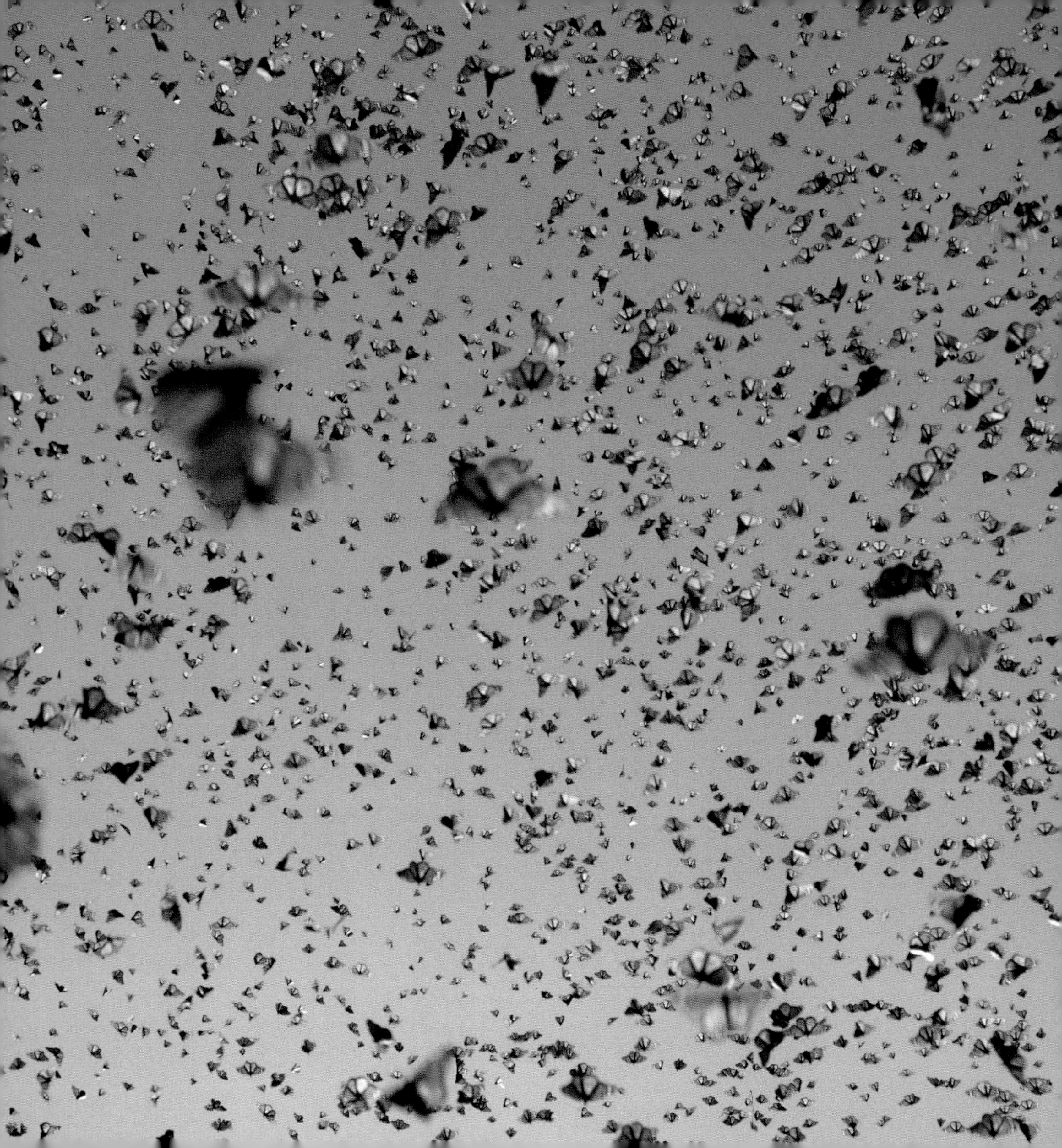

ACKNOWLEDGEMENTS

The body of knowledge from which I drew details for this book was built by many scientists. In particular, I am indebted to the pioneering work of Fred Urquhart and Lincoln Brower. Dr. Brower also read the manuscript and made many kind suggestions for its improvement.

I would like to thank Rob Sanders for inviting me to write this book and Candace Savage for discussion and encouragement. Nancy Flight oversaw the project and Anne Norman provided attentive editing and invaluable guidance and advice throughout.

FACING PAGE *Monarch butterflies gather in spectacular numbers at an overwintering site following their fall migration. Although not the only butterflies to migrate, they are the champions of long-distance insect flight.* FRANS LANTING/MINDEN PICTURES

INTRODUCTION

A warm September afternoon in southern Ontario, 1975, and I was driving east along Lakeshore Boulevard towards downtown Toronto. There were still three or four hours before sunset. The lake with its cheery sailboats was on my right, and ahead of me the black and grey skyscrapers of the city flashed sunlight from their mirrored faces into my eyes. As usual on a sunny fall weekend, joggers and cyclists dodged strolling families along the narrow strip of park between the roadway and thc lake. Bur not everything on this day was as usual.

FACING PAGE *Butterflies are specialized to feed on flower nectar, which is about 90 per cent water and 10 per cent sugar. This monarch drinks its liquid meal from a wild aster.* ROBERT MCCAW

Glancing through the side window, I saw an airborne stream of fluttering orange and black shapes wafting from the city, following the broad, gentle curve of the water's edge. At first, I thought these drifting scraps of colour were flurries of early fall leaves snatched from branches by a persistent lakeside wind. Moments later, when one of them skimmed against my windshield, I realized they were monarch butterflies (*Danaus plexippus*), travelling south on their famous migration, which I had heard about but never before seen.

I pulled over and got out of my car to savour the experience. The living clouds of brightly coloured insects bestowed a brief, sublime spectacle on surprised citizens, transforming the mundane, weekly pulse of the city to their far older rhythm. As I stood on the grass, butterflies flapped overhead and swerved around me, tempting me to think I could reach out my hand and scoop some up. But their speed and agility were more than a match for me, or for the laughing children who ran and jumped in futile pursuit of their beautiful, elusive prey. People paused and smiled in wonder as the monarchs flowed past traffic and pedestrians and trees, urged onward by the shoreline towards the southwest horizon.

Some of the butterflies settled briefly to rest on branches, soaking up warmth from the lowering sun before moving on. Many others, straying too low to the road, met an abrupt end to their journey. Struck down in their hundreds by passing traffic, their bodies blew against the curb and piled up in dusty heaps.

I picked up a dead butterfly to look at more closely, getting some of its powdery coloured scales stuck to my fingers. The multitude of tiny overlapping scales that cover their wings not only give butterflies their brilliant colours, they also supply the name of the insect order in which butterflies and moths are classified. The name—Lepidoptera—comes from the Greek words meaning "scaly wings."

The butterflies that flew past me on the outskirts of Toronto that day had another 4000 kilometres (2500 miles) of flying ahead of them. Those that survived the journey would eventually join tens of millions of other monarchs in mountainside forests west of Mexico City, where they would spend the winter before starting back north the next spring.

As it happens, the year 1975 and the city of Toronto are both significant in any account of the monarch butterfly. It was 1975 when the first spectacular overwintering site of the monarch butterflies in Mexico was revealed to the world at large, discovered after a search of nearly forty years. And one of the scientists whose years of research helped lead to the discovery was Dr. Fred Urquhart, a zoologist working at the University of Toronto.

The lifelong study of monarchs was a true vocation for Fred Urquhart, who began collecting butterflies as a child. In the charming preface to his classic book *The Monarch Butterfly,* he writes: "There were many factors involved in the study of these beautiful creatures which I found appealing. The search for caterpillars led to many a thrilling adventure as I combed the woods, examining every growing plant from lichens to tall trees for a discovery that would add a new species to my collection. Rearing the caterpillars provided an absorbing occupation at home, where my collection of rearing bottles teetered precariously on a row of rickety shelves attached to our garden fence."

His language evokes a long tradition of human fascination with butterflies, which have the distinction of being one of the few insects universally loved and respected. Inoffensive and pretty creatures, butterflies are absolved from the unpleasant associations of their relatives—the stinging, biting, buzzing, or disease-spreading kind. I have known only one instance in my life of a person deliberately killing a butterfly, and the sight was so shocking I vividly remember it thirty years later. When a teenager in England, I had a summer job as a bus conductor. A red admiral butterfly had flown into the bus and was fluttering along the aisles and bumping against the windows. Before I could usher it gently out, a woman irritably squashed it against the window with her hand. I could not have been more dumbfounded if I had seen someone whipping an angel with a cat-o'-nine-tails.

This book describes the life of one of the most remarkable members of the butterfly world. It also introduces a few of the people who have spent, in some cases, most of their lives trying to understand monarch butterflies, lured from one tantalizing question to another. Many things about these insects remain a mystery despite decades of research. No one knows, for example, how migrating monarchs find their way to their destination over such long distances.

To witness the flight of hundreds of thousands of butterflies as I did on that memorable fall day was a privilege—one I shared with many others along the monarchs' path through the United States and Mexico. It was a link, also, to observers of the past, who saw such sights long before our countries were created. Whether people not yet born will continue to see the monarchs' journey is a matter in the hands of the present generations.

Reluctant to fly in wet weather, a monarch butterfly shelters among pine needles until the rain passes and the sun comes out. GARY VESTAL

PART I

THE LIFE OF A MONARCH

Of the twenty-eight major orders of insects, the Lepidoptera (butterflies and moths) are surpassed in number of species only by the Coleoptera (beetles). Butterflies are more conspicuous than most moths, but much less numerous, accounting for only about 20,000 species of the 200,000 or so Lepidoptera known worldwide. Butterflies and moths are specialized nectar feeders, their mouth parts modified to form a long proboscis for drawing up the energy-rich food that flowers provide. They evolved in tandem with the huge increase in flowering plants about 65 million years ago.

FACING PAGE *Its outstretched wings span the breadth of a hand, making the monarch one of the largest North American butterflies.* KEVIN T. KARLSON

The North American monarch butterfly (*Danaus plexippus*) is part of a large family of butterflies found around the world, mainly in tropical and subtropical regions. Members of this family are abundant, large, and slow-flying, with striking wing patterns that make them among the most conspicuous of butterflies. The monarch undoubtedly reached North America from the neotropics, where there are still many non-migratory populations of various sizes, including those on the Caribbean islands, southern Mexico, Central America, and South America. During the last century, naturalists recorded the spread of monarch butterflies from North America to both the east and west, presumably carried by winds, by flight, or on ships. Stray individuals are occasionally found in western Europe today, and monarchs have established populations in Australia, in New Zealand, and on other South Pacific islands where milkweed plants (the main food of the caterpillars) have been introduced.

The vast majority of monarch butterflies in North America are migratory. These can be seen across the continent each summer, from coast to coast and from southern Canada to Mexico. They are split into two large populations by the Rocky Mountain Range, with eastern monarchs outnumbering western ones by at least a hundred to one. Although the monarchs on both sides of the Rockies resemble each other in appearance and habits, the mountains isolate the two populations from one another and there is probably little or no interbreeding among them. Each winter, all eastern monarchs migrate to a small number of sites in central Mexico, while the western monarchs overwinter at numerous locations along the coast of California.

One of the most intriguing features in the lives of all Lepidoptera is the dramatic metamorphosis that creates a butterfly from a caterpillar, or larva. The change is controlled by a balance of three different hormones, which cause the caterpillar's body to almost completely break down before the material is rebuilt into the adult insect. This trick of dividing their lives into two totally different forms has a great advantage. It means that adults and larvae can live in different habitats, feed on different foods, specialize in different tasks—in short, have twice as many opportunities to exploit their environment. Like Dr. Jekyll and Mr. Hyde, the monarch butterfly and caterpillar show two contrasting sides of the same life. One is an elegant and gregarious beauty with a penchant for courtship and travel, the other, a stay-at-home glutton with voracious habits and a solitary disposition. They seem almost to be two different species, their life history a symbol of resurrection.

FACING PAGE *The milkweed plant's large seed pods contain silky fibres that were once used to stuff life preservers. Monarchs do not usually lay eggs on such mature plants but prefer younger plants.* ROBERT MCCAW

TOP *A jewel-like egg nestles among the hairs on a milkweed leaf. Developing inside is a tiny larva that will eventually turn into a monarch butterfly.* MAVIS LESSITER/ HEDGEHOG HOUSE

BOTTOM *No bigger than a raindrop, a newborn monarch larva perches on its old egg case above a field of tiny leaf hairs.* PAT LOUIS

THE EGG

A day after mating, the female monarch searches for a good spot to place her eggs. Her choice of nursery is crucial. The tiny caterpillars that will chew their way out of the dome-shaped eggs in a few days' time will need food and shelter from the weather and enemies if they are to survive. And if they do not survive, her life's work will have been for nothing.

The monarch butterfly uses only one type of plant on which to lay eggs: the milkweed. Her larvae will eat nothing but the leaves of these tall, herbaceous perennials, which grow in clumps beside roadways and on abandoned farmland and other open areas, reaching heights of 120 to 150 centimetres (4 to 5 feet) during the summer.

Typically, an egg-laying monarch moves in a meandering, low-flying course over a likely area, skimming to within a finger's length of the tallest plants. Flying slowly against the breeze, she ignores the larger, flowering plants among the ranks of milkweed growing near the boundary of a field. They are old and tough and will soon be dying down after producing the large seed pods that complete their cycle. Like them, the butterfly is intent on establishing a new generation, and for this she prefers the small, tender leaves of young plants, with their promise of a longer future to sustain her offspring. She looks for healthy plants, shunning any with yellowing leaves that signal viral disease and rejecting those infested by milkweed bugs. She avoids as well those already taken by other monarch mothers, since their earlier-hatching larvae will be liable to view her eggs more as food than as fellow insects.

Hovering over a likely candidate with her legs outstretched, the searching butterfly touches a leaf with her feet. Like other insects, she has taste sensors on her feet, as well as smell receptors on her antennae. She uses her sense of smell to find milkweed plants at a distance, her sense of taste when she touches down. She seems at first to have only four legs instead of the insects' standard six, but a closer look shows the first pair, much shorter than the others, folded and held close to her body.

After settling on a plant and judging it suitable for her task, the gravid monarch grasps the edges of a leaf using two claw-like toes on the tips of her feet, and then curls her abdomen beneath and squeezes out a single egg. Most often, eggs are laid on the undersurface of a leaf, concealed for maximum protection close to the midrib near the base, but eggs can also be found on top of leaves, or in flower buds. She anchors the tiny capsule in place with a dab of gummy secretion and leaves it to its fate.

The egg is a ridged dome of creamy yellow, its colour blending with the pale tones of the young milkweed leaf. Inconspicuous as it is to most people, the monarch butterfly's egg inspires rapturous odes from lepidopterists, who have called it "one of the most exquisite objects in nature," and a "priceless gem cut by a master craftsman." A slight depression at the top of the egg marks the site of a minute pore through which air and moisture can pass.

After laying an egg, the monarch flutters away to find another plant, leaving only one egg per leaf to avoid future sibling rivalry (or cannibalism). Over the course of three or four weeks, she may lay a total of up to five hundred eggs, alternating intervals of egg laying with periods of feeding. The large number of eggs laid by each female helps ensure that at least some will hatch successfully.

FACING PAGE *Two caterpillars of different ages share a leaf. Over a period of about three weeks, the ever-eating larvae grow from the size of a pinhead to the length of a matchstick.* J.A. WILKINSON/VALAN PHOTOS

THE LARVA

Within the serene canopy of the egg, cells are multiplying furiously to build a miniature caterpillar. This process takes an average of five days to complete, the actual time varying with temperature. In hot weather, with temperatures over 26°C (79°F), the eggs may hatch in three days. If it stays below 18°C (64°F), they may take as long as ten days. Over the course of this period, the top of the egg slowly changes colour to a dark grey as the head of the larva develops inside.

The tiny caterpillar starts life as it will continue, chewing its way out of the egg and then consuming the rest of the eggshell before turning its busy jaws to the leaf beneath its feet. Only 2 millimetres (0.08 inches) long when it emerges, so small it could not escape drowning in a raindrop, the newborn larva browses at first on the fine leaf hairs growing on the undersurface of the leaf. But soon it is able to nibble away the more substantial material of the leaf blade itself, chewing out ever-widening holes from the place where it was born. A streamlined, juvenile feeding machine, the butterfly larva spends its days bulking up, building the proteins that will carry the insect through its nectar-sipping adulthood. In no more than three weeks, a monarch larva will increase its length by 25 times and its weight by about 3000 times.

In Latin, the word *larva* means "spirit, ghost, or mask." It's as if this small, wriggling creature that presumes to be the offspring of a beautiful butterfly is a fake, a mischievous imposter hiding its true nature until the time, a few weeks from now, when the crawling worm will finally drop its mask and rematerialize as the genuine child of its jewel-winged parents.

Like every insect, the monarch caterpillar has a tough cuticle enclosing its body. Constructed in segments like a suit of armour, the cuticle allows the caterpillar to move, while protecting the soft tissues beneath from damage and preventing its body from drying out. The cuticle, however, does not allow the caterpillar much room for growth, since the hard material from which it is made is not very stretchable. As the insatiable larva chews its way through leaf after leaf, its body swells and slowly pushes out every fold and wrinkle of the encasing cuticle until it becomes taut and smooth. At this point, like the Incredible Hulk, the caterpillar must literally burst out of its skin to get any bigger.

A monarch larva sheds its skin a total of five times. The periods between moults are known as instars, and last from two to seven days, with later instars being longer than early ones. When ready to moult, the caterpillar stops feeding and finds a secluded spot on the plant. There is a lot going on in the larva's small body at this time. Hormones secreted by cells in the brain cause a moulting fluid to dissolve the inner layers of the old cuticle, separating it from the epidermis. Freed in this way, the cells of the unattached epidermis can now divide to form a new, larger skin, which is folded like a concertina to allow for future expansion.

After the new skin has grown, the moulting fluid is reabsorbed, and material from the inner surface of the old cuticle is reused to build a new cuticle on top of the epidermis. When this is done, the caterpillar is ready to shed what is left of its old armour. The larva blows itself up by swallowing air and increasing fluid pressure in its abdomen, causing the redundant cuticle to split along lines of weakness built into it.

Graduating to the next instar in a fresh outfit, the larva crawls out of the old cuticle head first, then rests until the new outer layer hardens. Once ready to resume feeding, it begins by eating the old skin. With little protein in the larva's regular diet, the skin provides valuable materials for building new tissues.

The moult is a dangerous time in a caterpillar's life. Vulnerable when changing suits, some larvae get stuck during the process. Still struggling to free itself from its old skin, a trapped larva can attract the attention of predators, or die from exhaustion. For older, larger caterpillars, the moult may take as long as five to ten hours.

In its second instar, the caterpillar moves onto other leaves on the milkweed plant. Its shiny black head becomes patterned, and its pale greyish body develops conspicuous bands of yellow, black, and white, which become sharper and bolder with each subsequent moult. For most of the time, it feeds on the lighter-coloured undersurface of the leaf, where, at a distance, the bright bands that seem so gaudy close up blend into the background of shadowed foliage. With each passing day, the caterpillar grows plumper and more sluggish, more liable to attract the unwelcome attention of an insect-eating predator.

Chemicals from the milkweed plant make the monarch caterpillar's flesh distasteful to most animals, a subject described in more detail in Part 3. The larva's first line of defence, though, reduces the chance that a predator will ever get its jaws close enough to sample the flesh. At a sign of danger, such as the sudden vibration caused by a landing bird, the caterpillar immediately curls up and drops to the ground, lying quiet and hidden among the surrounding tall weeds and grass. Playing possum is an effective strategy. Even if a potential predator spots the larva falling, it would be unlikely to go to all the trouble of hunting for such a small meal. After remaining still for a minute or two, so as not to give away its location, the fallen caterpillar moves quickly away through the miniature jungle.

After falling from its eatery, the monarch caterpillar must get back onto a milkweed plant or starve. Curiously, this is a task for which it does not seem well equipped, lacking the sense organs needed to identify milkweeds at a distance. Monarch larvae have six pairs of simple eyes, which are good only for detecting movement and light intensity. They can sense vibrations (a form of hearing) through their skin, and they have a well-developed sense of taste. But caterpillars must be in contact with a milkweed before they can tell what it is.

Observing that the larvae knocked off a plant crawl away in all directions, monarch expert Fred Urquhart carried out an experiment. He placed a fresh milkweed leaf on a large sheet of paper, and then released larvae of various instars at different distances from the leaf. Unless they were placed facing the leaf, and less than 5 to 7 centimetres (2 to 3 inches) away, the larvae appeared to find it only by chance, after wandering about at random. Of course they may have been disconcerted by the unfamiliar surface, but observations in the wild showed the same pattern: grounded larvae are apparently oblivious to milkweed plants growing within 30 centimetres (1 foot) of their crawling search, and depend mainly on luck to return them to their sole source of food.

Being stranded on terra firma holds risks other than that of starving, as the larva can now fall prey to many small insect-eating ground-dwellers, such as ants and mice. In this predicament, the mother butterfly's choice of plants for her eggs could be crucial for the larva's survival. Most monarch butterflies lay their eggs on milkweeds growing in clumps, rather than leaving them on solitary plants growing apart from others. Whatever their reason for doing this, the habit helps their offspring, since a caterpillar on the ground among a group of milkweeds has a much better chance of getting quickly back to safety.

By the time the fifth and final instar comes to an end, a monarch caterpillar will have chewed its way through an estimated 10 grams (a third of an ounce) or so of milkweed leaves. With, say, 50 million other monarch caterpillars on the continent, a single generation converts about 500 tonnes of milkweed into monarch matter.

Monarchs typically go through three to five generations during the course of a summer, depending on the weather and how far north they live. Their growth rate varies with the temperature, the length of daylight, and the quality of the milkweed, and caterpillars born early in the season (in May or June) or at the end of the season (in September) develop more slowly than those born in mid-summer.

Nearly three weeks after hatching from its egg, the caterpillar takes its last meal. Moulting hormones are once again flowing through its body and it must look for a sheltered place to shed its skin. Crawling to the ground, the caterpillar expels its last waste and moves off on the urgent quest. The next moult will not produce a larger caterpillar but a chrysalis: the mummy-like crucible in which the caterpillar's tissues break down and reorganize to create a dramatic metamorphosis into a butterfly.

Its growing days behind it, a monarch caterpillar prepares to turn into a pupa. Shortly after suspending itself upside-down like this, the caterpillar will split its skin for the last time. GAVRIEL JECAN/ART WOLFE, INC.

THE PUPA

The technical term for the next stage of the life cycle is *pupa,* a Latin word meaning "doll." The pupa's inert form conjures up the image of a sleeping baby wrapped in swaddling clothes. Like a baby, a pupa is vulnerable. A windblown leaf or stem striking it during the hours before its soft body hardens could puncture it and kill the pupa, or deform the butterfly developing inside. To avoid such hazards, caterpillars typically choose a well-protected site for their final moult, somewhere sturdy and safe from wind and rain, such as below the overhang of a fence rail, or the angle where a thick tree branch joins the trunk. Before settling down to moult, a caterpillar stops and waves its head around in a wide circle to determine that the spot is free from immediate risk.

Like all caterpillars, the monarch larva has a silk-making gland inside its body; the liquid silk is drawn out through a structure called the spinneret on its lower lip as the larva spins a small, round button of silk fibres on the underside of the fence rail or branch. Having laid down this anchor, the caterpillar rests for an hour before gripping the silk button with small, curved spines on its anal claspers, wriggling its rear end back and forth to entangle the hooks and get a secure hold. Then, like a trapeze artist, the caterpillar lowers its body so it is hanging from its silk anchor head down, but curving slightly up to form the letter *J.*

The pupa is already forming within the larval skin, as a new hormonal mix orchestrates the cellular rearrangement. The first outward signs are a fading of the flamboyant yellow bands, giving way to a dull, translucent blue-green. Time now to cast off the last caterpillar cloak. The pupating caterpillar pumps fluid into its thorax, swelling its body and splitting its skin for the fifth and final time. The expanding front end pushes the shrinking skin back along the suspended pupa and up to the silk button.

The pupa must now step out of its old skin without releasing its grip and falling. To do this, it uses a structure called a cremaster or pupa stalk, furnished with small, club-like spines at the tip. While still suspended by its shrivelled skin, the pupa pulls the cremaster free and pushes the spines into the silk button until they are firmly embedded. Having switched to the new holdfast, the pupa spins rapidly around, loosening its old grip and shaking off the unwanted skin. The entire process from the making of the silk button to the shaking off of the larval skin may take nearly twenty-four hours.

For the next fifteen hours the pupa is soft, bluish, and grub-like in appearance, with deep furrows marking the outlines of the butterfly-to-be. The cuticle slowly hardens and darkens over this time, and a number of shiny gold spots appear on its surface. Golden

FACING PAGE *A newly formed pupa is soft and furrowed. It hangs like a ripening fruit from its black cremaster, or pupa stalk, anchored in a button of silk.* DAN WRAY / FIRST LIGHT

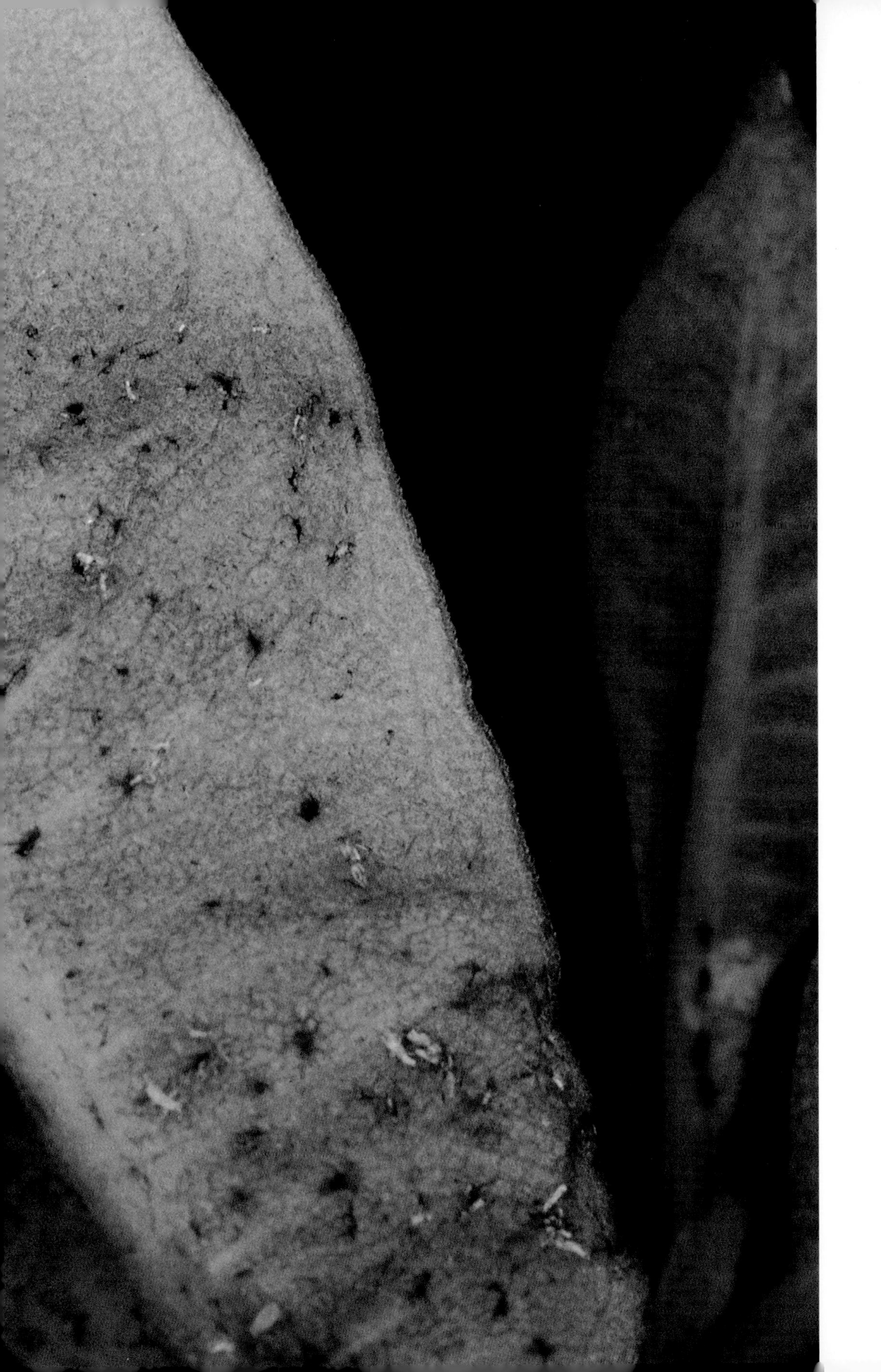

As the pupa hardens, it turns a beautiful green colour with gold dots.
OLA JENNERSTEN/NATURFOTOGRAFERNAS BILDBYRÅ

markings are characteristic of many butterfly pupae, giving this stage of the life cycle its other name of *chrysalis*—a word taken from the Greek term for "gold." Whether these gold spots have any function or not, nobody yet knows. In the past, some scientists have speculated that they act like light meters, monitoring the passage of night and day, or that they control colour development in the wings, but there is no evidence for these ideas. It may be that the gold spots are simply a by-product of the metabolic changes going on inside the insect, or they may help disguise the chrysalis.

The metamorphosing monarch remains entombed for about a week, a motionless and enigmatic entity between the crawling larva and the restless adult. Most people probably imagine the transition going on in the pupa as a steady merging from one form to the other, with a part-caterpillar, part-butterfly stage halfway, like the computerized morphing effect shown in some rock videos and science fiction movies. In fact, during the chrysalis stage, the caterpillar is almost completely broken down into a soup of cells before the butterfly is built up, and a peek inside the pupa case after a couple of days would reveal a liquid mess rather than a chimera.

The change of form and function affects every part of the insect's being, from its senses to the way it moves and feeds. Buds of tissue in the thorax grow and develop into wings. The larva's leaf-nibbling jaws dissolve and new adult mouth parts grow, later fitting together to make a hollow tube through which the adult butterfly will draw nectar. The long intestine shrinks to match the new diet, and sex organs appear for the first time. Long, delicate antennae develop on the insect's head, and the twelve simple eyes of the caterpillar are replaced by the two huge compound eyes of the adult. All these changes are finely co-ordinated, so none comes too soon or too late.

A darkening of the green chrysalis signals that the butterfly is getting ready to emerge. Within a day after the colour change, the adult's reddish-orange wings are clearly visible through the thin pupa casing, and the chrysalis twitches and twists. The metamorphosis is almost complete, and the insect will soon enter the last stage of its life as a monarch butterfly.

FACING PAGE *The monarch butterfly's distinctive black and orange wings show through the translucent skin of the pupa case a few hours before the insect is ready to emerge.* J. SHAW STACK/FIRST LIGHT

PAGE 24 *Little more than a week ago it was a caterpillar. Now its amazing metamorphosis is nearly complete. A butterfly gets ready to meet the world with new sense organs, mouth parts, reproductive organs, and wings.* FRANS LANTING/MINDEN PICTURES

PAGE 25 *The butterfly clings to its old pupa case for an hour or two after emerging. Before it can fly, its wings must expand and harden.* GAY BUMGARNER

Odour receptors on the antennae, together with taste receptors on its feet, guide the butterfly to a source of food. The antennae droop downward as the insect draws nectar through its long tubular proboscis. JEFF FOOTT

THE BUTTERFLY

The convulsive movements of the chrysalis continue for a few moments until fine cracks appear in the old pupal skin behind the head. More jerking opens up the fractures, which lengthen and join together, letting a jagged corner of the now-transparent covering plate curl back to show part of the hairy black body beneath.

The butterfly chooses the timing of its debut carefully. A bright, calm day just before noon gives it the crucial time it needs to complete the transformation and harden its wings while conditions are still warm. Waving long legs through the widening split, the butterfly begins to slowly drag itself free of the chrysalis case, making a woebegone figure with its swollen abdomen and unexpanded wings. The newly emerged insect clambers up the old case and clings on limply, its wings hanging vertically down. If the butterfly falls at this point, or cuts its wing veins on a sharp edge, the wings may become twisted or distorted. They still need a couple more hours to expand and dry in the sun and air, in order to become the thin but tough structures capable of carrying the butterfly over long distances.

For the next fifteen minutes, the butterfly's abdomen pulsates as it pumps fluid into the branching tubes that run through each of its four wings. As it does this, the abdomen grows smaller and the wings grow larger. They are not expanding like balloons, but uncrinkling as fluid flows into them. When fully stretched and taut, the monarch butterfly's wings span about 10 centimetres (4 inches) and display the beautiful black, orange, and white pattern that makes this species unmistakable.

With its wings dry and stiff, the butterfly can fly if disturbed. Often, however, it will remain clinging to its perch for the rest of the day and overnight, giving its body and new sensory system time to become fully attuned to its surroundings. When the sun's rays warm its body early the next morning, the monarch opens and closes its wings a few times, and then drops from its resting place and flutters off to explore.

No matter how familiar it becomes, the sight of a butterfly breaking from its dark tomb and taking flight remains a powerful symbol of life—not just of birth, but of rebirth, of resurrection. In the Greek language, "butterfly" and "soul" share the same word: *psyche.* Classical art portrays the soul as a beautiful young woman with butterfly wings, and in the book *Metamorphoses*, the Roman writer Apuleius tells how the mythical princess Psyche captured the heart of Cupid, the god of love.

For two or three days after the monarch re-enters the world as a butterfly, it spends its days (weather permitting) visiting blossoms. For the rest of its life, the butterfly will feed only on nectar, water, or fruit juice, a diet that can sustain astonishing feats of aerial endurance. Freed from an earthbound existence to range over a larger world, the adult insect has keener eyes than the larva, able to see a wide spectrum of colours, including ultraviolet light. The butterfly finds flowers by sight, and uses sensitive odour receptors on its antennae to judge their qualities as it flies closer. Taste sensors on the feet come into play as soon as it settles.

Butterflies may seem frail and aimless as they flutter jauntily from one flower to another, but they are neither. Try to catch a butterfly and you'll soon see they are alert, strong, and agile fliers. A monarch's wings flap five to twelve times a second, depending on the flight mode. They sometimes glide, sometimes cruise—at about 18 kilometres (11 miles) per hour—and, when alarmed, speed away at up to 50 kilometres (30 miles) per hour. Monarchs can survive a lot of battering, and can fly even with as much as half their wing area gone. They usually fly only by day, and shelter under leaves when it rains. On very hot and sunny days, they may retreat to the shade to avoid overheating.

COURTSHIP AND MATING

A few days after emerging from the chrysalis, monarch butterflies begin the vital business of courtship and mating. Throughout the summer, males and females can be seen together rising sharply upward in spiral courtship flights over meadows and other open areas. Males are the faster fliers, and relentlessly chase after any female they spot in their neighbourhood, losing the race only if she darts for cover into the foliage. The courtship ritual is an intricate set of displays and responses that can last several minutes, but it may be aborted by a distracting gust of wind or a passing bird at any point until the male grips the female, leaving the participants to begin another chase. Matings can occur at any time of day but are most frequent in the late afternoon.

Male monarchs can be told apart from females not only by their behaviour but by their appearance. The male monarch has slightly thinner black markings along his wing veins than the female, and he has scent pouches that look like small black spots, one on each of his hind wings. The pouches contain specialized scales that are thought to absorb sex pheremones produced by hairpencil glands at the tip of the male's abdomen. In the closely related queen butterflies (*Danaus gilippus*), sex pheremones are essential for successful courtship, but in monarchs, the males can catch and mate with females even after their hairpencil glands have been removed.

Following a wild pursuit flight, in which the male may brush the female's head and antennae with his hairpencils, the male butterfly pounces on the female and carries her to the ground. Snapping his wings open and shut, like some eighteenth-century courtier with a large, ornate fan, the male monarch pins the female down and bends his abdomen sideways to grasp her securely with his pincer-like abdominal claspers. If disturbed at this point, the couple may fly off, still linked together by his grip. After settling again, he transfers a package of sperm to her through a short, needle-like organ. Copulation usually occurs in a secluded area of dense shrubbery and can last from two to fourteen hours. The female retains the sperm sac inside her body until needed, releasing sperm for fertilizing each egg just before it is laid. Both males and females may mate with several partners, and females may carry several sperm sacs.

There are equal numbers of males and females when monarch caterpillars hatch from their eggs, but if you go butterfly hunting in a field one summer's day you are likely to catch more of one sex than the other. The reason for this is that male and female butterflies march to the beat of a different drummer and tend to be found in different places at different times. Males are inclined to pass the days flitting around flowering plants, while

FACING PAGE *After a vigorous pursuit through the air, a male captures and carries a female to a secluded spot and encloses her with his wings.* ROBERT MCCAW

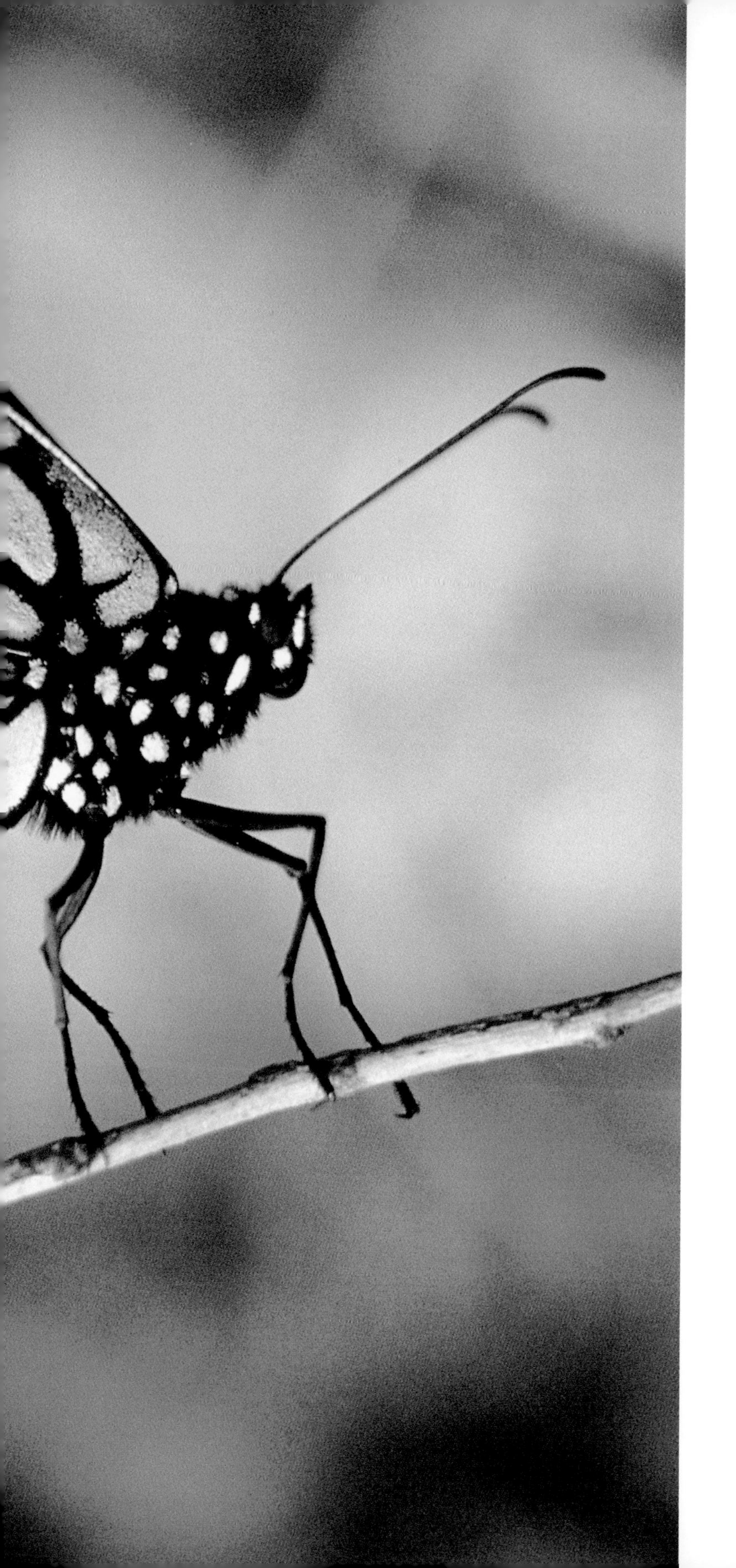

Mating butterflies may remain linked together for several hours. Mating is frequent over spring and summer, both sexes pairing with several partners. In this photograph, you can clearly see why monarchs are sometimes called "four-footed butterflies."
FRANS LANTING/MINDEN PICTURES

females are obliged to spend much of their time hunting for patches of milkweed on which to lay their eggs.

During the summer months, monarch butterflies live for two to six weeks, dying shortly after mating and egg laying is completed. The generation of monarchs that crawl from their chrysalis cases under the September sun, however, is destined for a life dramatically different from that of its parents and grandparents. These butterflies haven't the time or inclination to court, and mate, and lay eggs. Within hours of their first flight, they feel another compulsion, equally important for the survival of the species. Driven by the more immediate need to save themselves from the lethal hand of winter, they will soon escape to warmer latitudes, postponing their nuptial rites to another season. Their life span, if they survive the hazards of migration, may be as long as eight months.

In this classical sculpture in the Louvre, Paris, Psyche is shown with wings. The word psyche *in Greek means both "butterfly" and "soul."*
ALINARI/ART RESOURCE, NY

The changing season triggers a new behaviour in the butterflies that emerge in fall. Instead of mating, they will migrate south, flying as far as 130 km (80 miles) in a day.
ROBERT McCAW

PART 2

MIGRATION

As the path of the planet begins to shade the northern pole from the sun for the next six months, the summer playgrounds of the monarch butterflies start to close down. The shorter, cooler days of early fall signal butterflies and other animals of the north to take shelter or migrate, calling creatures as diverse as hummingbirds and gray whales to turn their faces south.

Many insects solve the problem of cold temperatures and lack of food by slowing down cellular activity, overwintering in a dormant state with the flame of life set low. Some butterflies endure these hostile months as eggs,

FACING PAGE *"A living, breathing, palpitating picture ..." is how the local newspaper described a massive swarm of monarchs that migrated through Cleveland, Ohio, in 1892, alarming some of the residents. This flock is at a winter roost in Mexico.*
GARY VESTAL

some as larvae, some as pupae, and some (such as the mourning cloak and red admiral) as adults. Prisoners of the thermometer, staying at about the same temperature as their surroundings, hibernating butterflies must choose their sleeping quarters carefully, or this slumber may be their last. The survivors will be those that find nooks to protect them all winter from the prying eyes of predators and the probing fingers of polar winds.

For monarch butterflies, falling temperatures signal a change in behaviour. The unhurried gliding from flower to flower that traced out the warm days of August gives way to earnest flapping southward on the sharper September air. Their bodies must be warm to work, and these creatures of the sun could be immobilized if they don't get going, snagged by one of the cold fronts beginning to creep down from the Arctic.

FACING PAGE *Migrating monarchs are reluctant to fly over large bodies of water. This butterfly waits on the north shore of Lake Ontario for good flying weather before continuing its southward journey.* ROBERT MCCAW

The two black spots on its hind wings identify this handsome specimen as a male monarch. The spots are scent pouches that store sex pheremones used to attract females. FRANS LANTING/MINDEN PICTURES

THE JOURNEY SOUTH

At the northern limit of their range in Canada, monarch butterflies begin to move out on their migration as early as August. The peak of the exodus comes in mid-September, and by the end of October all are gone from the area north of the Great Lakes. They keep ahead of the withering fall flowers, following a trail of blossoms to Mexico. Sugar-rich nectar from goldenrod and aster, the stored energy of summer's light, turns to heat in their flight muscles as it boosts them along their way.

It is during this fall migration that monarch butterflies are at their most conspicuous to casual observers, passing in steady procession during the day, or settling to roost at night in sheltered trees and bushes. On clear, warm days, they fly with a slow gliding movement interspersed by sudden bursts of speed. Strong fliers though they are, the weather remains an ever-present threat to their lives, and monarchs avoid taking to the air when there are strong winds or a good chance of rain.

They seem to be good forecasters, as Fred Urquhart relates in *The Monarch Butterfly*. Describing a gathering of over a thousand monarchs that he found roosting on a tree one morning in September 1957, he records their reluctance to move off despite a warm temperature and slight breeze. When Urquhart disturbed the butterflies, they flew a short distance, circled round, and settled back on the branches. Although human observers saw no signs of change in the weather, the insects must have known better. A few hours later, their unwillingness to abandon the overnight shelter was vindicated when the sky grew overcast, winds picked up to speeds of 40 kilometres (25 miles) per hour, and thunderstorms swept the area. Urquhart speculated that the butterflies might be able to sense the falling air pressure that comes before a storm.

Large gatherings of butterflies build up at points along the migration route as a combined result of weather and topography, when their carefree flight paths over farmland, cities, woods, and hills are brought to stern discipline by the coastline and shores of large lakes. Hesitating to leave the sight of land, they follow the water's edge. Channelled together by the forbidding barriers, thousands of monarchs from the north coalesce into growing flocks, which tumble into butterfly bottlenecks on peninsulas and promontories, fingers of land pointing their way across the waves. Clustered by the shore, the insects may be held in place for days by cold, damp, or wind, their numbers swelling as others join them. These traffic jams produce well-known sites for butterfly watchers, who flock every fall to places such as Point Pelee and Long Point on the north shore of Lake Erie to see the gathered monarchs waiting for suitable conditions to brave the flight over open water.

FACING PAGE *Clustered on flowers at summer's end, monarchs build up fat reserves for their long migration flight.* JIM BRANDENBURG/ MINDEN PICTURES

Monarch butterflies have been recorded by sailors many kilometres from shore on the Great Lakes and off the Atlantic Coast of North America. Occasional individuals have even been found on the west coast of Great Britain, having presumably survived unscheduled diversions across the ocean. Carried by winds, there is little they can do once launched from land, and most probably die as a result of being blown off course.

Travelling only by day, the migrating butterflies can be found each evening in small groups, or ones and twos, alighting on bushes and trees beside fields and bordering ravines and streams. Before the deepening twilight eclipses their bright colours, they must find a safe roosting place for the night. It's not so much that they need to rest. The problem is that nighttime temperatures may dip too low for sustained flight. If caught in the open and forced to the ground by cool conditions, they run the risk of being captured by small nocturnal hunters, such as beetles, ants, mice, shrews, or toads.

The butterflies settle on the downwind side of trees and close their wings. Using small hooks on their legs to get a good grip on a bit of corrugated bark or the serrated edge of a leaf, they hang motionless until dawn, secure from being dislodged by a stray gust. In the morning, as the temperature climbs, they open their wings and slowly drift away, one after another, until the tree is bare.

The monarchs sweep south in flurries like advancing weather systems, cyclones of butterflies heading for the tropics to merge and build into a hurricane of hundreds of millions of beating wings over Mexico. Taking advantage of rising warm air currents, they may circle to altitudes as high as 1200 metres (4000 feet) before gliding down to catch the next wave up. They are returning to their ancient homeland, the centre from which the species spread out many thousands of years ago, for monarchs are really tropical butterflies that have expanded their range.

Scientists speculate that monarchs may have first begun migrating from their original home in Central and South America with the start of the Pleistocene Era, about 2 million years ago, when a changing climate and evolutionary adaptations allowed milkweeds to spread north. At the same time as the monarchs followed the milkweeds to more temperate latitudes, drier winters developed in Mexico. The plants on which the butterflies and caterpillars depend withered with the dry winter season, and the monarchs could survive this period in Mexico only in a dormant state. The combination of shifting climate and vegetation, it is thought, produced the pattern we see today, where monarchs take advantage of the abundant food in the United States and Canada each summer but must hurry back to Mexico to avoid freezing each winter.

FACING PAGE *At staging posts along their migration routes in North America, monarchs settle each night in sheltered trees to wait out the cold night air. On Long Island, New York, butterflies open their wings to catch the morning sun before continuing their long journey south.* KEVIN T. KARLSON

PATHS OF FALL MIGRATIONS

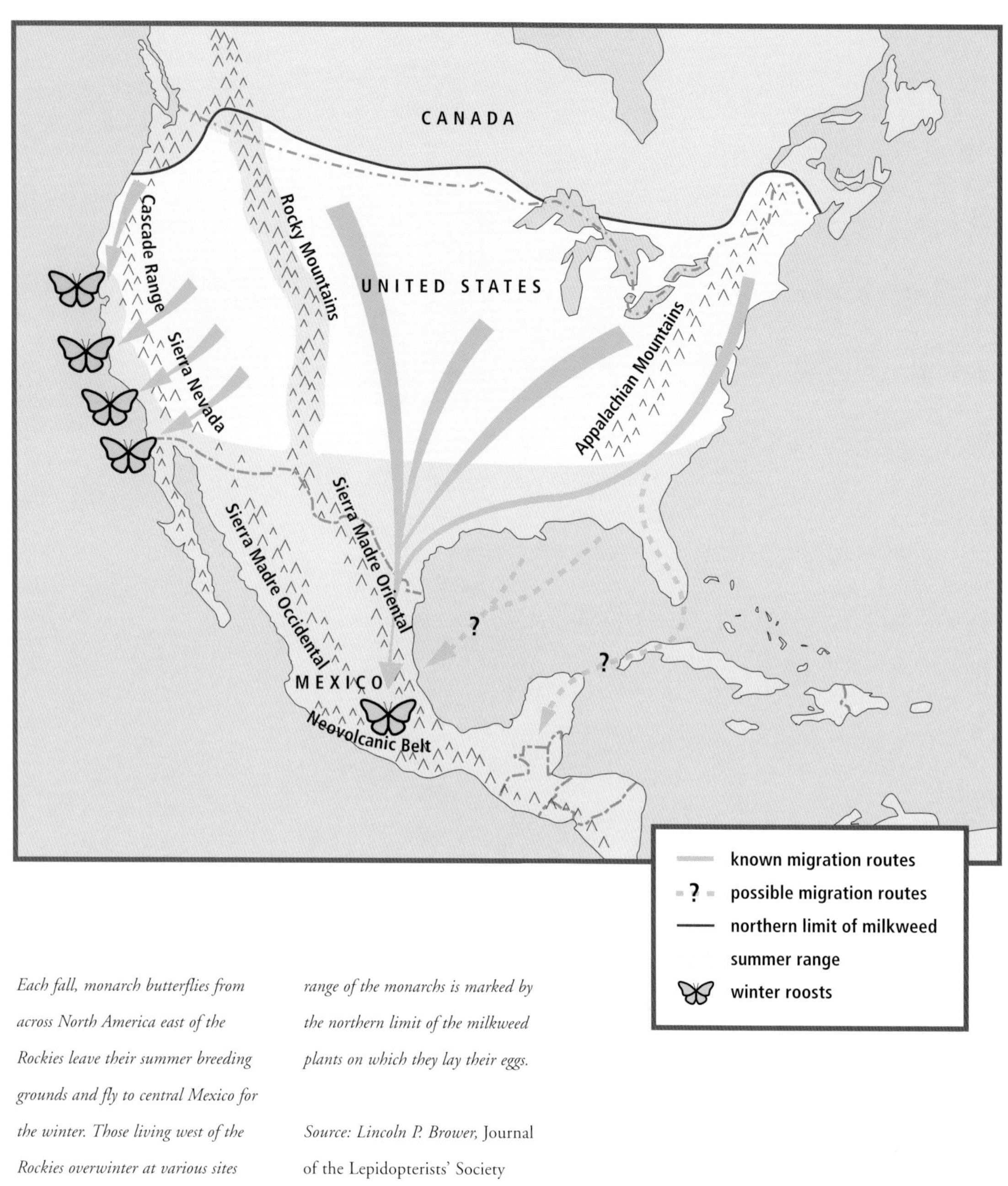

Each fall, monarch butterflies from across North America east of the Rockies leave their summer breeding grounds and fly to central Mexico for the winter. Those living west of the Rockies overwinter at various sites along the Pacific Coast. The northern range of the monarchs is marked by the northern limit of the milkweed plants on which they lay their eggs.

Source: Lincoln P. Brower, Journal of the Lepidopterists' Society *49, no. 4 (1995): 322.*

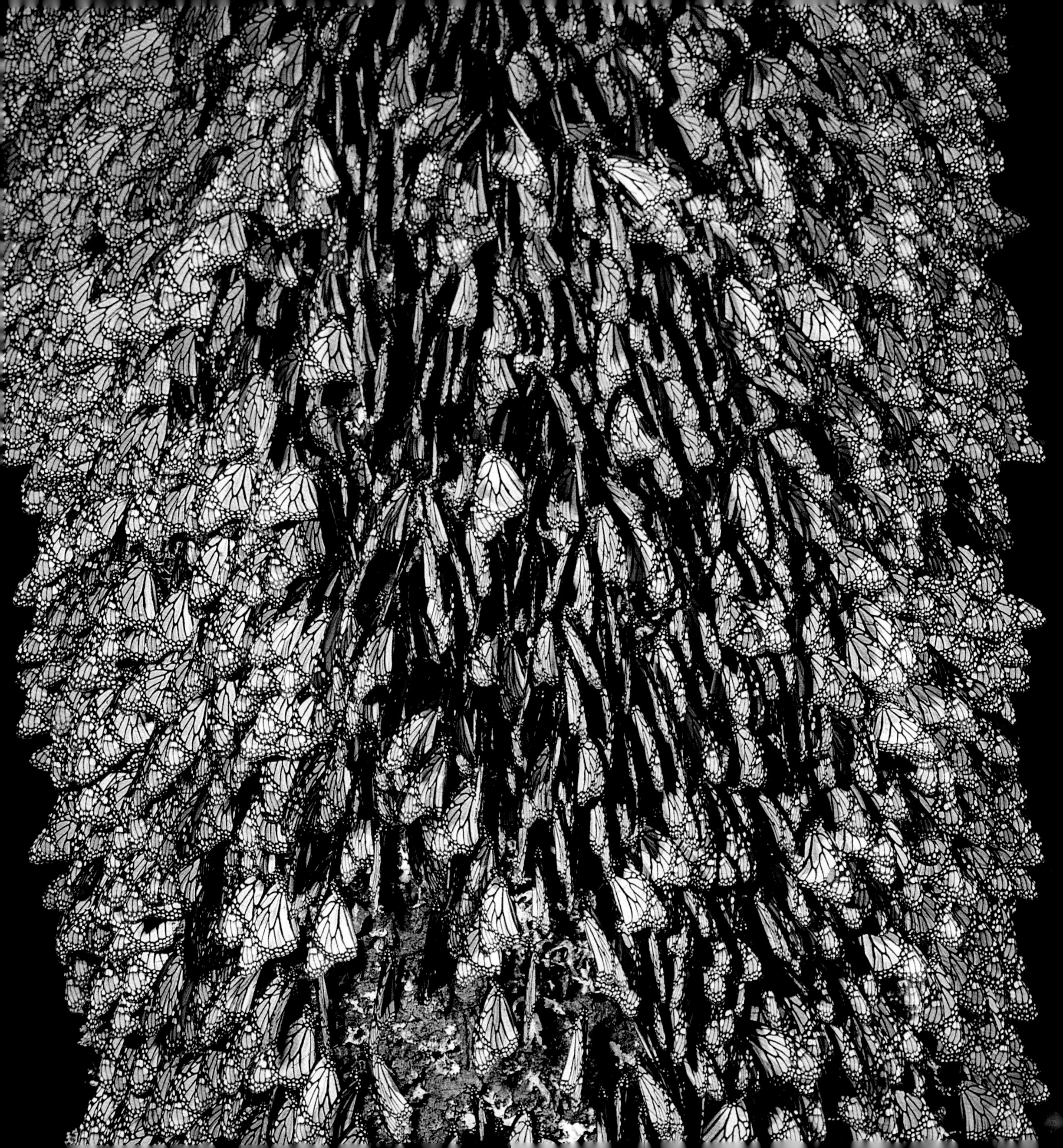

TRACKING THE MIGRANTS

The monarchs' disappearance from eastern North America every fall is both a familiar event for millions of people and a source of many mysteries for scientists. When Fred Urquhart began his study of monarchs in 1935, one of the biggest puzzles was, Where do all the butterflies go? It took forty years of painstaking research to find the answer.

Big, bright, and abundant as they are, monarch butterflies are impossible to follow overland for very long as they may fly six hours and cover 130 kilometres (80 miles) each day on their migration flights. Observers watching butterflies flutter by could estimate numbers and record flight speeds and directions, but in a matter of hours the skies would be empty, the butterflies' destinations still unknown. What scientists needed was something like the method used to study the movements of migrating birds, which had been tracked for years by means of numbered bands fastened around their legs. But how do you mark individual butterflies so their identities can be reported back by whoever finds them?

The obvious answer was wing tags, but the technical difficulties of finding tags that work were another matter. Tags had to be lightweight, waterproof, adhesive, easy to attach, able to carry legible information that would not fade or wash off, and, of course, not interfere with the monarchs' flight. After twenty years of trying different methods that failed, Fred Urquhart and his wife, Nora, who had joined him in his research, finally found a workable tag in the 1950s, based on the self-adhesive price tags used on jars in grocery stores. About the size of a thumbnail, each tag was printed with a number and mailing address where the finder could return it. They were attached by folding them over the leading edge of the butterflies' right front wing and pressing gently between thumb and finger.

That was only the beginning. The Urquharts needed volunteers to tag monarchs in different parts of the continent and to be on the lookout for tagged insects. They wrote articles in naturalist magazines, and the word soon spread. A network of monarch taggers and watchers grew, and results began to come in as people discovered individual butterflies hundreds of kilometres from the places where they were caught and marked.

By 1964, a few hundred volunteers had tagged about 70,000 monarchs. From the picture building up on his map of tag sightings, Urquhart knew that Mexico was a likely destination for the migrants, but he still did not know whether they stopped there or went farther on to Central America, and it was not known whether they spread themselves over the country or congregated in a few small areas. When the stunning discovery of the first winter roost gave the answers in 1975, a total of more than 400,000 monarchs had been tagged by thousands of volunteers over twenty years.

PREVIOUS PAGE *Dense clusters of monarchs cling to a tree trunk at an overwintering site in Mexico, their wings folded tight against the cold air. They remain in a semi-dormant state from December to March, surviving on fat reserves stored in their abdomens.* GARY VESTAL

FACING PAGE *Labelled wing tags help scientists and butterfly watchers track the migration movements of monarch butterflies from Canada and the northern United States to Mexico.* PETER ARNOLD/PETER ARNOLD, INC.

73266

WINTER ROOSTS IN MEXICO

The bare summit of Cerro Pelón in south-central Mexico stands at 4 kilometres (2.5 miles) above sea level, capping the dense stands of oyamel fir forest that cover the steep slopes below. Remote and inaccessible until recently, this volcanic mountain range west of Mexico City gave up its secret on January 2, 1975.

An American member of Urquhart's volunteer team, Ken Brugger, with his Mexican wife, Cathy, and their dog, Kola, had spent the previous year venturing up nearby mountain trails in their spare time looking for signs of monarchs. Everyone knew there must be monarchs in Mexico, because all their flight paths from the southern United States led there, but then the butterflies seemed to vanish. Searching for monarchs during a long drive through Mexico in the winter of 1968, the Urquharts had seen none.

The Bruggers got help from local woodcutters who knew about the wintering sites, and, just after New Year, they were climbing the side of Cerro Pelón when they spotted a monarch butterfly spiralling above them. Going up in the direction it flew, they began to see dead butterflies on the forest floor, and shortly afterwards their path took them into a thick grove of oyamel close to the summit. Adjusting their eyes to the dark, they saw flashes of orange in patches of sunlight that penetrated the canopy, and realized they were looking at tens of millions of monarchs festooning every tree trunk and branch. The clusters of butterflies were so incredibly thick that heavy branches sagged under their weight.

The winter roosts of monarch butterflies in Mexico are among the greatest biological finds of the century, a zoological equivalent of the tomb of Tutankhamen, a treasure trove of history and beauty combined. Thanks to the additional sleuthing work of monarch expert Lincoln Brower and his colleagues, another dozen overwintering sites were found within a few years after the Bruggers' first discovery, all in thick oyamel forests, all at altitudes of about 3 kilometres (2 miles) above sea level, and all within a radius of 80 kilometres (50 miles) of Cerro Pelón.

The numbers and density of butterflies in these sites are truly astonishing. Census figures in different years give population estimates from 60 million to 90 million, congregating in small sites with a total area no greater than a few hectares. One site alone, measuring only 100 metres (330 feet) in diameter, was reported to contain 30 million butterflies. Incredible as the numbers are, it must be remembered that the butterflies in these roosts must be only a percentage of all those that originally set out on the journey south in the fall, many of which succumb to cold, exhaustion, accidents, drowning, starvation, or other mishaps.

FACING PAGE *Most North American monarchs spend the winter among these mountain peaks in Mexico. They congregate in only a handful of relict fir forests that provide the conditions they need to survive the cold weather.* FRANS LANTING/MINDEN PICTURES

Monarchs from all over North America east of the Rockies begin to arrive in these roosts in late November to early December, some having flown for three months on a journey of about 4000 kilometres (2500 miles). Depending on the weather, they go into a state of almost complete inactivity shortly after arriving, hanging motionless from the trees with their wings closed during the cool periods. On warm days, they may fly short distances to find water or nectar, but most have enough fat stored in their bodies to see them through the winter without feeding. The thick canopy protects them from wind, rain, and occasional snow, keeping them cool enough to remain dormant but not so cold that they freeze.

Towards the end of winter, the butterflies that flutter from the trees each day to find nectar are tattered and thin, their energy reserves depleted, their wings worn and torn. Most of those flying around the roosts at this time are males in poor condition, eager to mate if they can, as they will not survive much longer. The plumper butterflies of both sexes still clinging to trunks and branches are saving their fat reserves for the homeward flight.

FACING PAGE *Trees that harbour monarchs through the winter provide the precise conditions the insects need to survive. They must be protected from the wind, not too hot and not too cold, not too wet and not too dry.* FRANS LANTING/MINDEN PICTURES

WINTER ROOSTS IN CALIFORNIA

Researching the history of monarchs in California, Lincoln Brower found reports dating back to as far as the 1860s describing large clusters of butterflies gathering on pine trees each winter on the Monterey Peninsula, and swarms flying out of the area each spring. Other accounts detailed monarchs flying westwards in the fall from the mountains to the coast, in numbers some observers estimated to be in "countless millions."

By early this century, butterfly watchers had distinguished between the migratory movements of the eastern and western populations. Although the exact origins of the butterflies that moved to the mild climate of the Pacific Coast were unknown, it seemed likely from observations that they streamed down from breeding areas along the Sierra Nevada and the Cascade Range, and perhaps from other mountains and valleys extending east to the Rockies.

Researchers in the 1980s documented reports of at least 200 monarch winter roosting sites along the Pacific Coast, from north of San Francisco to Baja California in Mexico. Tagging studies helped establish some of the distances these western butterflies move, but the details of flight routes on both the fall and spring movements are still largely unknown. Monarchs tagged during winter in the Monterey area have been recaptured in spring in the Sierra Nevada foothills, and some tagged individuals flew north and east from their winter sites as far as 465 kilometres (290 miles).

Like their counterparts in Mexico, the western monarchs depend for their overwinter survival on particular conditions of the microclimate. Both the high-altitude oyamel forests of Mexico and the sea-level clumps of Monterey pine and eucalyptus along the California coast provide the fine balance they need. Sheltered among the thick, needle-bearing branches of the trees each winter, the insects are cold enough to remain torpid but not so cold as to become frozen. It stays warm enough for them to maintain their clusters but not so warm that they become too active. And it is wet enough to protect them from the risk of forest fires or drying out but not so wet that they become chilled. These conditions cannot easily be duplicated, and any threat to the trees is a threat to the monarchs.

FACING PAGE *A warm winter's day in California encourages a group of overwintering monarchs in a eucalyptus grove to stir themselves. In the early afternoon, clusters of butterflies may burst from the trees in an explosion of colour and fly a short distance to feed.* MICHAEL EVAN SEWELL

THE JOURNEY NORTH

From about the second week in March, the monarchs that overwintered in Mexico begin the return journey to their breeding grounds in the United States and Canada. Flying along river valleys and mountain passes, they move north keeping pace with the warming weather and the blooming flowers. Feeding and mating continue along the route, and mated females begin to lay eggs as soon as they find suitable patches of fresh milkweed. Compared with the fall flight south, these butterflies are in more of a hurry, flying low and racing to the milkweed before they run out of energy.

Most of the returning butterflies lay their eggs on milkweed in the southern states and along the Gulf Coast and then die, leaving the next generation to continue the journey north as far as Canada. Although a few individuals may have the endurance to make the complete round trip, the majority of monarch butterflies that fly into the northern states and Canada later in the summer are individuals born along the migration route. It is not known whether particular northbound butterflies use the same flight paths they took to Mexico the previous fall, or whether monarchs leaving the winter roosts fan out at random.

FACING PAGE *A floral fuel station attracts a small cluster of monarchs. Small sensors at the tip of the proboscis help guide it to the food, and a pumping mechanism in the butterfly's head draws the liquid into a simple digestive system.* JOHANN SCHUMACHER/PETER ARNOLD, INC.

THE SEARCH CONTINUES

Far from closing the book on monarch research, the discovery of the winter roosts in Mexico stimulated new interest in these fascinating and beautiful insects, triggering both more discoveries and more questions. In the early 1990s, a network of monarch mavens across Canada, the United States, and Mexico came into their own by joining forces on the Internet—a technology tailor-made for tracking butterfly movements and sharing information about their behaviour.

An Internet site named Monarch Watch links fans of the butterflies with regular bulletins of sightings and research news. After students in Mexico post messages in March to report monarchs leaving their roosts, regular visitors to the site can follow the unfolding story of the migration through the ensuing weeks as observers from all over the continent send in news of what's happening in their region. Monarchs just arrived here. Monarchs seen flying north there. Smaller numbers than usual. First caterpillars. Hot weather. Few milkweeds. And so the modern version of the bush telegraph keeps everyone up to date on the butterflies' progress.

Monarch Watch originated in 1993 as an outreach program of the University of Kansas. Started by entomologist Orley R. "Chip" Taylor as a research project, it has become an overwhelming success with thousands of students, teachers, volunteers, and other interested people, quickly adding science education and conservation to its original goal of data collection.

The monitoring program works in conjunction with tagging. In the first year of its launch, Taylor sent out 48,000 of the tiny green glued labels, mainly to school groups. Three years later, 80,000 tags were winging their way over the continent, and Taylor estimated that there were at least 30,000 students in thirty states who had already tagged and studied monarchs through this program.

As a result of their experience, Taylor and his team updated the tagging method that had been used for over thirty years since it was first developed by Fred Urquhart. Introduced in 1995, the new technique is simpler and quicker, with less risk of harming the butterflies. Instead of folding an adhesive tag over the leading wing edge, the new, smaller tags are pressed gently against a dab of glue applied to the hind wing. Kits sent out with the tags explain the details of tagging and the proper way to record observations for research. Taggers usually do more than simply tag. They also report dates, directions of flight, weather, the sex of the butterflies, and other useful details.

Monarch Watch truly captures the imagination of students, many of whom also rear their own butterflies and discover the excitement of being part of a real, hands-on research project. As pioneer researcher Fred Urquhart realized long ago, we cannot understand the details of monarch migration without recruiting a lot of observers.

After over forty years of research on one of the most conspicuous and common animals in North America, and with new information pouring in from monarch watchers every year, there are more questions than ever before still waiting for answers. Many details of the butterfly's basic biology and its migration remain a mystery. How do they know when to begin migration and where to fly? How do they find their way over many hundreds of kilometres to a tiny area in Mexico? What produces the difference between a non-migrating monarch born in summer and a migrating one born in fall? What proportion of monarchs survive the journey?

Modern technology is giving scientists new tools and new clues. For example, researchers working with University of Florida zoologist Lincoln Brower developed a method of "fingerprinting" monarchs by analyzing the milkweed chemicals in their bodies. Each species of milkweed has a different chemical composition, detectable in the tissues of the larvae that feed on them. These chemical fingerprints persist in the adults, allowing scientists to determine the area in which a captured butterfly began its life. This evidence helped confirm that the first monarchs to arrive in Canada each summer hatched from eggs in the southern states, and did not originate either from Mexico or from the northern states.

Genetic analysis of monarch butterflies from different locations has also turned up some interesting and unexpected evidence about their movements and distribution. During the summer, separated breeding populations in different regions east of the Rockies may develop small genetic differences from one another over a few generations. When the monarchs gather at their winter roosts in Mexico, however, buttterflies from all regions mix together. The small variations become swamped among so many millions of butterflies and are not statistically distinguishable. The monarchs do not appear to stay together with others from their region in the huge roosts, or when they return north. In early spring, a monarch that had flown down from, say, Ontario might mate with one that originated in Arkansas. Overlapping generations add to the genetic mixing. For example, eggs laid in the Gulf states by the first monarchs to return north in spring could produce butterflies a month later that then mate with late-leaving monarchs en route from Mexico. In effect, all the monarch butterflies living east of the Rockies form a single, vast, interbreeding megapopulation.

What about the monarchs living west of the Rockies? A more recent genetic technique, developed since the advent of genetic engineering in the mid-1970s, gives an even more startling picture. Comparing the mitochondrial DNA of monarch butterflies from the eastern and western United States, as well as from Mexico and the West Indies, scientists

Chemical analysis of a monarch's tissues tells scientists what species of milkweeds the animal fed on as a larva. Such tests, together with genetic studies, have provided new evidence about the pattern of monarch migration across the continent. ROBERT McCAW

A.V.Z. Brower and T.M. Boyce found almost no variation. Their report, published in the journal *Evolution* in 1991, points out that this level of similarity in the DNA from geographically isolated populations is dramatically different from nearly all other groups of animals studied.

Since there is no evidence from observations or tagging that monarch butterflies cross the Rocky Mountains to interbreed, why are the different populations so genetically uniform? Brower and Boyce argue that the most plausible explanation is that all of today's monarch butterflies have descended fairly recently from a very small number of individuals. Before the eastern and western migration patterns developed, they suggest, there was a population bottleneck with very few breeding females. Not enough time has yet gone by for their descendants to develop significant genetic changes from their common ancestors. (A situation like this is found among cheetahs, which at one point in the past must have come very close to extinction. Cheetahs throughout Africa today are genetically very similar, creating problems of inbreeding.)

From field records to DNA analysis, researchers are adding to our store of knowledge about monarchs every year. But the information they gather gives us more than a growing database on these remarkable butterflies and their world. It is only from such painstaking and continuous record keeping that we can learn in time if monarch butterflies are in danger. Are their numbers across North America going up or down from year to year? Early signs of declining populations could be vital warning of unhealthy changes in the environment that are harming butterflies—and perhaps people as well.

FACING PAGE *Torn wings often are mementos of a close encounter with a bird's beak. The monarch's strong flying ability seems little affected by such misfortunes.* GAVRIEL JECAN/ART WOLFE, INC.

PART 3

THE MONARCHS' WORLD

The numbers and distribution of monarch butterflies across North America each summer can vary dramatically from year to year. Some fall seasons produce vast clouds of migrating butterflies, while in other years they are rarely seen. It is only since the winter roosts were discovered, giving us a fuller picture of the butterflies' range and annual cycle, that we have had a broad enough picture to begin understanding why.

The size of the first generation of monarchs to fly into Canada in spring depends on how many butterflies survived the previous winter in Mexico,

FACING PAGE *As the first light of dawn brightens a monarch's treetop roost, the butterfly prepares to set off on the day's business.* RON WATTS/ FIRST LIGHT

and on the breeding success of an earlier generation in the Gulf Coast states as the monarchs spread up the continent. Survival at these different stages depends in turn on the interaction of such things as weather, predation, disease, and the abundance of plants on which the butterflies and caterpillars depend.

As well as the annual changes in monarch numbers, there is evidence that their populations have undergone huge historical shifts since the last century. From his research of archival records, Lincoln Brower makes the case that the abundance of monarch butterflies in various regions of North America today is mainly the result of human activities over the past 150 years. These activities have made large areas of their traditional breeding grounds unsuitable for monarch butterflies, while opening up some new areas where they were once scarce.

Following the last Ice Age, about 10,000 years ago, the Great Plains of the continent's heart were clothed with prairie grasses, wildflowers, and no less than twenty-two different species of milkweed. It was here that most monarchs made their summer home, until pioneers of the 1800s moved into the fertile plains, dug in their plows, and converted most of the natural prairie to fields of grain. By the 1880s, there was little left of the plants on which monarchs depend, and they were lucky, in a way, that the same period that saw the decline of the grasslands also saw the decline of the great deciduous forests in the east.

Millions of hectares of forest throughout the eastern United States and around the Great Lakes region were felled during the last half of the nineteenth century and beginning of the twentieth, creating cleared areas into which common milkweed (*Asclepias syriaca*), goldenrod, thistles, and other weeds rapidly spread. For the monarchs, it was a case of "you win some, you lose some," as the population centre moved from the prairies to the east. This historically cleared forest region, with its thousands of small farms, became one of the main breeding areas of monarch butterflies in North America, and remains so today.

A fallow field, unused by farmers, makes an ideal butterfly habitat with its crop of wildflowers. The abundance and distribution of butterflies depends on the availability of flowers like these. GAY BUMGARNER

Like a vast, throbbing cloud, the monarch population expands and contracts in a yearly cycle. From these condensed colonies in the mountains of Mexico, the butterfly clusters will unfold next spring and spread across North America. GARY VESTAL

MONARCHS AND MILKWEEDS

The distribution of monarchs is tied closely to the distribution of milkweeds, and the intimate link between butterfly and plant is of a kind that has evolved often during the long, entwined history of insects and flowering plants. Rooted to the spot, unable to run from their enemies, plants have invested much of their evolutionary energy developing defences against organisms that would feed on them. Their specialty is chemical warfare. Past masters at making malign molecules, plants produce a huge variety of poisons, feeding inhibitors, emetics, narcotics, and hormones designed to deter animals from nibbling at their tender parts. The zesty tastes of herbs and spices we use to titillate our palates are in many cases intended to make insects throw up or go somewhere else.

The milkweed is one plant that punishes browsers. Its thick, white sap contains a lethal brew of cardenolides (heart poisons), which produce vomiting in low doses and death in higher doses. Most animals shun milkweed, but monarchs are one of the few exceptions. Their tissues are insensitive to the chemicals' effect on the nervous system, and monarch caterpillars thrive on the leaves that others leave alone.

There are over one hundred species of the sturdy, upright, perennial herbs growing in North America, all members of the genus *Asclepias.* Different species vary in their levels and composition of deterrent chemicals, and younger leaves on a plant contain more than older ones. Feeding caterpillars often pinch the stem of a leaf with their jaws before nibbling the blade, causing the leaf to droop and wilt. They do this to cut off the flow of sticky latex that carries the poisonous cardenolides into the leaves.

Having evolved a physiological strategy that lets it feed on milkweed without harm, the monarch pays a price. It is a specialist, unable to feed on anything else, and it can live and reproduce only where milkweeds grow. That is not a problem as long as milkweeds are common and widespread, but it makes the monarch vulnerable to their disappearance.

On the plus side, the monarch caterpillar has few competitors for its only source of food, and can dine with plenty of elbow room. More importantly, monarchs turn a liability into an asset by using the plants' feeding deterrent for their own protection. The caterpillars sequester cardenolides from their diet in their bodies, building up concentrations even higher than those found in the plant. If grabbed by an animal, the caterpillar regurgitates a noxious cardenolide-containing fluid, which quickly causes the attacker to drop it.

The protective chemicals accumulated by the caterpillar remain in it through metamorphosis, giving the nectar-sipping butterfly the same distasteful tissues. Observations

FACING PAGE *Milkweed plants are regarded as undesirable weeds in many parts of North America because of their toxic sap. But as they are the sole food of monarch caterpillars, the fate of these plants is intimately connected with that of the butterflies.*
ROBERT MCCAW

and experiments confirm that monarch butterflies are unpalatable to most animals, which either avoid them altogether or quickly reject them and vomit after an initial taste.

It's clearly better to be left alone than to be sampled, and most inedible animals advertise their unpleasantness as part of their defence. Both monarch caterpillars and butterflies have clear, bold markings and colours that help predators learn quickly that they are not potential meals. An inexperienced bird needs only one serious bout of vomiting to remember not to look for seconds, and after that probably regards a monarch butterfly much as a dog might regard a skunk.

FACING PAGE *Bright orange butterflies flitting through a meadow of yellow goldenrod make a pretty fall picture, but insect-eating birds are no admirers of the colourful monarchs. The unpleasant taste of these conspicuous butterflies makes them all but predator-proof.* JAMES L. AMOS/ PETER ARNOLD, INC.

MIMICRY

The monarch butterfly's flamboyant markings are a celebrated example not only of warning patterns but also of another classical ecological interaction: mimicry. A different and unrelated species of North American butterfly, the viceroy (*Limenitis archippus*), looks very similar to the monarch, and is believed to have evolved its copycat colours as a survival strategy, taking advantage of the protection given by the monarch's appearance. It is only the adult stages of the two species that look alike. The eggs, larvae, and pupae of the monarch and viceroy are quite different, and the two species feed on different plants. A third look-alike, the queen butterfly (*Danaus gilippus*), is a close relative in the same genus as the monarch.

For decades, the monarch and the viceroy have been used in textbooks to illustrate mimicry. But evidence for the theory of mimicry has more often been assumed than tested. While there are many examples of it throughout the animal kingdom, the case of the monarch-viceroy has been disputed for almost as long as it has been written about.

The concept of mimicry originated in the Victorian era, during the heyday of biologist-explorers. Taxonomy was a mania of the times, and many collectors searched remote regions for samples of exotic animals and plants to sort and classify into groups. Pursuing butterflies and other insects under the canopy of the Amazon rainforest during the 1850s, Henry W. Bates, an English naturalist, was struck by the fact that some butterflies from different groups looked alike. In 1861, two years after the publication of Darwin's *On the Origin of Species,* Bates published his observations on protective mimicry as evidence of natural selection at work.

Bates's name lives on in the theory of Batesian mimicry. The bold colours and patterns on certain animals, he reasoned, must give them some benefit and help them survive. A pattern evolves initially because of the advantages of advertising a disagreeable or dangerous quality. It is a warning signal. But after such a pattern has developed and become effective, other species, without dangerous qualities, may exploit it for themselves. They avoid predators by looking dangerous, just as some species avoid predators by looking like a leaf or a twig. The original unpalatable species is called a "model" and the harmless and palatable species that resembles it is called a "mimic."

There are a couple of necessary concomitants to this scenario. The mimics must live in the same area as their models. And there are usually fewer mimics than models. If mimics are too numerous, the bluff becomes less effective. There will be a higher chance that palatable mimics get eaten, running the risk that predators learn to misread the sign for "danger" as "dinner."

FACING PAGE *This viceroy butterfly at first looks like a monarch, but it is slightly smaller. The two species are well-known examples of mimicry, a survival strategy common among tropical butterflies, which helps protect them from predators.* ROBERT MCCAW

The queen butterfly is a look-alike relative of the monarch found in the southern United States, the West Indies, Central America, and South America. Feeding tests with birds suggest it is much more palatable.
GARY VESTAL

For years, it was generally assumed that because viceroys do not eat milkweeds they are palatable Batesian mimics. Experiments in the 1950s had shown that blue jays reject monarchs as distasteful, and are equally reluctant to eat viceroys, but amazingly enough the experiments needed to actually compare the palatability of viceroys and monarchs were not carried out until the 1990s.

In a paper published in *Nature* in 1991, bluntly titled "The Viceroy Butterfly Is Not a Batesian Mimic," David Ritland and Lincoln Brower put a dent in the neat picture. The scientists offered red-winged blackbirds an assortment of viceroys, monarchs, and queens. To prevent the birds distinguishing between insects from their appearance (rather than their taste) they provided only the abdomens of the butterflies. As a control, they also included abdomens of species known to be palatable. To their surprise, the researchers found that viceroys were as unwelcome to the birds as monarchs. Only about 40 per cent of both species were eaten, compared with 70 per cent of queens and 98 per cent of controls.

The viceroy, then, is not a harmless mimic, but a noxious insect with a warning pattern in its own right. But all is not lost. The viceroy and monarch still resemble each other. Both are mimics of a different sort, showing a type of mimicry that was also first described in South America in the last century.

Fritz Müller was a German lepidopterist who, like Bates, did his butterfly collecting along the Amazon. Describing his "remarkable case of mimicry in butterflies" in 1878, Müller argued thus: if every species of unpleasant-tasting butterfly develops its own distinct warning pattern, predators of butterflies will have a lot of learning to do—at the cost of many mutilated or dead individuals of each species of prey. There is a selective advantage if two or more species of poisonous butterflies living in the same area adopt a generic pattern, sharing the benefits and lowering the risks. Because predators have only one warning pattern to learn, fewer butterflies get sacrificed as tests. This type of pooling, or mutual resemblance, is called Müllerian mimicry.

Unlike Batesian mimicry, there is not a model and a mimic in Müllerian mimicry, but a converging evolution to a common warning pattern. There is also no restriction on the numbers of each species, as there is with Batesian mimicry. The more the merrier. Abundant species still benefit from sharing their pattern, while scarcer species gain even greater advantage.

Mimicry turns out to be a fascinating and complex subject, not nearly as simple as it seems at first glance. It is particularly common among tropical butterflies, and some species even have different appearances in different areas of their range, depending on which other species share the area. There are many ecological and evolutionary lessons still to be learned from more field studies and experiments.

PREDATORS, PARASITES, AND DISEASE

Warning colours and poisonous chemicals work as a defence against predators only if the predator hunts by sight and is affected by the toxins. One of the surprises discovered at the monarchs' winter gatherings in Mexico is that large numbers of the butterflies there are eaten by birds and mice. Toxic and unpleasant though the butterflies are, it seems that such a bonanza of easy-to-catch protein is too tempting to ignore. Just as monarchs are able to circumvent the milkweeds' defence, so a few predators have evolved ways of safely making a meal of monarchs.

Thirty-seven species of birds live in the forests where monarchs overwinter in Mexico, but only two regularly eat the butterflies. They defy the monarchs' poison in two different ways. Black-backed orioles reduce their cardenolide intake by slitting open the butterflies' cuticles, where most of the poison is stored, and scooping out the soft contents of their abdomens. Black-headed grosbeaks have stronger constitutions. They seem able to tolerate moderate levels of cardenolides in their bodies, and eat monarch abdomens whole.

Both species have been seen feeding daily at monarch roosts in Mexico, visiting the sheltered butterfly groves in small flocks. Each bird can pick off from fifteen to forty butterflies per visit, so that, over an entire roosting season of three to four months, birds may be responsible for the deaths of millions of monarchs. While there's no evidence that predators have any impact on numbers of monarchs during the summer, these attacks at the roosts have been estimated to kill 10 per cent or more of the winter population at some sites.

Scientists have speculated that the monarchs' habit of clustering in vast numbers might have evolved, in part, as a defence against winter predation. Butterflies at the centre of a cluster are much less likely to be eaten than those at the edges, so the survivors are those with a tendency to be highly gregarious. The same selection pressure might also favour a tight synchrony in arrival at the overwintering sites. Late arrivals will be forced to settle in the vulnerable positions on the outsides of the group, while there is an advantage for getting to the roosts early and together.

Mice, too, take their share of the brightly coloured banquet. The black-eared mouse is a regular monarch predator, the only one of five species of mice living in the region to feed on the butterflies. In fact, it is the only mammal known to resist the monarchs' chemical defence. It avoids getting sick both by not eating the butterfly cuticles and by having some physiological tolerance of cardenolides in its body. Unlike the bird predators, mice take only butterflies near the ground, or freshly dead ones on the ground. An individual mouse might eat thirty to forty monarchs in a night.

The butterfly roosts help boost the mouse population. Scientists have observed that mice move into the roost area after the butterflies arrive, and they produce more young and are in better condition than black-eared mice outside the roost area, which do not use this food source. During the course of the winter, the mice might consume as much as 5 per cent of the butterflies in a roost.

The historical shift in monarch distribution from the prairies to the cleared eastern forests may have played a part in making the butterflies more vulnerable to predators. The dominant milkweed species growing in the east produce less toxic cardenolides than the native prairie species. Caterpillars feeding in these areas will contain the weaker poisons, which gradually disappear as they age and may make the adults coming from these regions more palatable.

Most of the facts and figures about monarch predators come from studies of animals that eat the adult butterflies. But equally important, though less is known about them, are the predators of caterpillars. The most important of these are not birds or mammals but insects, which sense the world in a different way. Many insects hunt by smell, and cardenolides may have a different effect on, say, wasps than on woodpeckers.

Ambush bugs and stink bugs have been seen killing and eating monarch larvae, and many caterpillars are threatened by dull-coloured flies called tachinids, whose larvae feed and grow inside the caterpillar's body. An adult tachinid fly lays its eggs on a monarch caterpillar, and the fly larvae that hatch out immediately burrow into the caterpillar and settle down to eat its tissues and fluids. They make the caterpillar weak but do not kill it until they are ready to leave. At that point, they inactivate the caterpillar by destroying part of its nervous system, then bore their way out, drop to the ground, and pupate.

An occasional parasite of monarch caterpillars is a small, slender member of the wasp order called a braconid fly. Using her long ovipositor, a female braconid deposits a single egg inside a caterpillar. The egg divides to produce as many as thirty-two separate embryos, which develop into genetically identical larvae. Like the tachinids, these larvae feed only on the non-essential tissues of the caterpillar at first, before killing it when they emerge. Researchers at Cornell University have recently begun studying the effects of cardenolides on these parasitic wasps. There is some evidence the wasps may be able to sense the concentration of toxic chemicals in the host's body and may selectively lay their eggs in caterpillars with lower levels of cardenolides.

One potentially important death-dealing parasitic protozoan (*Ophryocystis elektroscirrha*) appears to specialize in monarch and queen butterflies, as it has only been found on these two related species. The parasite is spread in the form of spores that cling to the wings of the butterflies. Infected adults are weakened by the protozoa developing inside them, and in this condition they are unlikely to survive the winter. Butterflies pass spores

FACING PAGE *A black and yellow garden spider finds a monarch in its web. Although some butterflies lose their lives to hazards like this, by far the biggest cause of unexpected death is the weather.* ROBERT MCCAW

from one to another during mating, and female butterflies shed spores from their wings onto milkweed leaves while laying eggs. The hatching larvae ingest the spores, carrying the infection into the next generation of butterflies.

Like every living thing, monarchs can succumb to a variety of diseases caused by microbes of one sort or another. Their sequestered plant chemicals, though, may give them some immunity to most fungal diseases, since both milkweed plants and monarchs appear naturally resistant to fungal infections. A viral disease of insects that occasionally kills monarch caterpillars, pupae, and butterflies is caused by the nuclear polyhedrosis virus. Infected animals become lethargic, lose their bright colours, and slowly turn brown and then black as the cells inside them are killed and their insides putrify and turn liquid.

By and large, predators, parasites, and diseases have only a relatively small effect on monarch numbers. The main cause of the wide fluctuations in population size appears to be the weather. Winter storms can kill millions of butterflies in their winter roosts, while the summer's regime of sunshine and showers largely determines the supply of milkweeds and nectar sources during their breeding season, producing either feast or famine. Subject to such natural hazards, monarchs are able to roll with the blows, bouncing back from lean years as soon as more favourable conditions allow. Human activities are now tilting that long-established balance against the butterflies, however, threatening to upset the monarch's world and posing a very real danger that the spectacular migrations may soon become a thing of the past.

FACING PAGE *Most stink bugs (also called shield bugs) feed on plants but some are carnivorous, such as this specimen dining on a monarch caterpillar. Stink bugs repel enemies with a strong-smelling fluid that can produce bad headaches in some people.*
ROBERT McCAW

PART 4

THE NEED FOR CONSERVATION

In late December 1995, a rare cold spell brought heavy snowfalls to the mountains of southcentral Mexico. For two days, an endless curtain of white flakes descended on the forests where a substantial part of the North American monarch butterfly population has its seasonal sanctuary. Newspapers reported as many as 20 million of the hapless insects had died in the storm, blown from their roosts onto frozen snowbanks under the falling snow, where thousands upon thousands of them were found by forest rangers who toured the area early in January.

FACING PAGE *A tree trunk is festooned in its gay winter cape of monarch butterflies. There is growing pressure to log forests in Mexico, threatening to remove the trees on which millions of butterflies depend for their survival.* RON WATTS / FIRST LIGHT

Luckily, those early estimates later turned out to be exaggerated. Many of the butterflies found on top of the snow had not been killed, as the first reports assumed. After the cold weather passed and the sun had warmed their bodies, they were able to revive and fly back onto the trees. A more accurate appraisal obtained later by Mexican researchers gave the storm's death toll at 4 million, or about 7 per cent of the population in the affected area. It was less than feared, but still a cause for serious concern.

News of the monarch's plight was broadcast across North America, jolting people into realizing that a unique natural wonder of the world was in danger of disappearing, little more than twenty years after it was first revealed. But the real threat to the monarchs in this region, experts agree, is not the weather alone. Butterflies have survived winters in these cold mountaintops for thousands of years. What puts their roosting sites under severe pressure today is logging.

FACING PAGE *The dazzling sight of this butterfly grove in Mexico may be denied to future generations if strong measures aren't taken today to protect the monarch and its habitat.* FRANS LANTING/MINDEN PICTURES

PAGES 90–91 *A study in contrasts, an insect of sunshine and flowers lies stranded in the grip of frost and snow. Overwintering monarchs can survive moderately cold temperatures and snowfalls if they are not wet, but damage to their forest sanctuaries leaves them vulnerable to rainfall and winter storms.* GARY VESTAL

THREATS TO WINTER ROOSTS

The oyamel forests in which the monarchs overwinter are specialized high-altitude ecosystems, found only on the higher peaks of volcanic mountain ranges in Mexico and occupying less than half of one per cent of the country's land area. They are relicts of the extensive fir forests that once covered this southern region during the last period of glaciation, only to be replaced by other species of trees after the climate began warming about 10,000 years ago. Isolated on their mountain slopes in little more than a dozen small islands of vegetation, these rare oyamel forests are the most vulnerable to deforestation of any type of forest in Mexico.

Recognizing the vital need to protect the monarchs' winter habitat, the Mexican government passed a law in 1986 creating five separate reserves in the roosting sites, covering a total of 160 square kilometres (62 square miles). Each reserve has a central zone where tree cutting is strictly prohibited, and a surrounding buffer zone set up to protect the area from further human interference. But despite the laws, logging goes on at a frightening rate in both protected and unprotected areas. As large areas of forest disappear completely, the butterflies are forced to concentrate even more in the few suitable places left.

The roots of the deforestation dilemma are poverty and a growing population, a pattern familiar throughout the tropics as farmers expand their fields from valleys onto the slopes of forested mountains, with disastrous consequences everywhere. Farmers living near the monarchs' roosting sites make their living, in part, by cutting trees for building materials and firewood, and forest regeneration is hampered by their cattle trampling and eating fir seedlings.

The butterflies cannot survive in any old clumps of trees. Needing the very precise conditions that only the oyamel forests provide to protect them from the winter weather, they are affected by even a moderate disturbance of these vulnerable ecosystems. The risk was outlined a few weeks after the 1995 storm in an article written for the *New York Times* by Lincoln Brower and Mexican poet and leading environmentalist Homero Aridjis: "When intact, a forest serves as an umbrella and blanket, protecting the butterflies from freezing rains. The logging creates gaps that allow rain and snow to fall through the forest canopy and onto the butterfly clusters. As the weather clears, the life-sustaining heat radiated from the butterflies' bodies leaks out through these holes in the blanket of trees and the monarchs freeze to death."

Studies show that rain-dampened butterflies are more susceptible to the cold, and the weather immediately after a storm may be even more important than the snow itself. Dry

FACING PAGE *Throughout North America, wildflower habitats that support butterflies and other insects are being reduced by herbicide sprays, mowing, and urban development.* JIM BRANDENBURG/MINDEN PICTURES

butterflies can survive long periods at temperatures near freezing, but logging leaves fewer areas with the dense canopies the butterflies depend on to keep from getting wet.

Logging is not the only hazard to monarch butterflies in their Mexican forest refuges. Air pollution from fires, lit by farmers to burn stubble and weeds from the fields, sends grey smoke drifting into the reserves. On cold days, the smoke causes butterflies to drop from their perches. On warm days, it sends them fluttering in frenzied flights that disrupt the structure and density of the colony. The clouds of smoke signal an unhappy message: Unless local residents can be involved in conservation programs, and their economic needs met in ways that do not threaten the trees, the butterflies' future there looks bleak. If these few forests are damaged, the entire pattern of the butterfly migration throughout North America is at risk. Alarmed by the seriousness of this threat, Lincoln Brower describes the monarch migration as an "endangered phenomenon."

FACING PAGE *Beautiful and vibrant even in death, an arrangement of butterfly wings on the forest floor resembles the delicate flowing patterns of Tiffany glass. Our world will have lost one of its most remarkable phenomena if we sacrifice those few sites where monarchs now gather in overwhelming numbers.* GARY VESTAL

THREATS TO SUMMER BREEDING AREAS

Although their winter roosts are the Achilles' heel of monarch butterflies, they face growing difficulties when dispersed throughout their summer homes as well. A high winter mortality might be quickly offset by a good breeding season, but a big blow in Mexico followed by poor conditions in the United States could be a one-two punch that knocks the monarch population out.

As the butterflies return north each spring, one of their first needs is a supply of milkweed on which to lay eggs and rebuild their population. But the plants that monarch caterpillars depend on for food are viewed in many parts of North America as weeds that landowners are legally obliged to destroy. In practice, milkweed control is a local matter, frequently neglected unless there are specific complaints. Milkweed was originally classified as a plant pariah due to its reported danger to livestock, and the official attitude remains, threatening to reduce the butterfly's already limited breeding range.

Changes in land use draw the noose around milkweed tighter. The main habitat for this plant in southern Ontario and Quebec, where a large part of the monarch population in Canada breeds, is on abandoned farms and roadside verges. Much of this farmland has been uncultivated since the 1960s as a result of the economic pressures that make it difficult for family farms to survive. While initially of benefit to the butterflies, the cleared land with milkweed transforms back into brush and woodland after a few years of neglect, shutting the insects out.

Throughout much of North America, many of the smaller farms that provide prime breeding grounds for monarchs are either being swallowed by large, intensive agricultural operations or encroached upon by sprawling communities. Big agribusiness has fewer niches for wild plants and uses more of the chemical sprays that kill both plants and insects. Suburban developments cover fields and hedgerows with a veneer of houses, roads, and malls. Caterpillar vehicles in construction sites replace caterpillars in meadows.

As well as the milkweeds needed by the caterpillars, adult monarchs need wildflowers to survive, floral fuel stations where they can top up with nectar for their flights. But these, too, have disappeared from many areas of North America, thanks to a combination of herbicide sprays, mowing, urban expansion, and pollution. In the shrinking and degraded habitat left to butterflies to make their living, the vicissitudes of the weather become a serious matter. Periods of drought or frost that stunt wild plants might once have been only a temporary setback in one region or another. Today, bad weather can spell an entire year's breeding failure over a large area of the continent.

FACING PAGE *A monarch caterpillar climbs the stem to a milkweed flower. Different species of milkweeds have different coloured flowers, some of them with a pleasant scent.*
D. CAVAGNARO / PETER ARNOLD, INC.

PAGES 98–99 *A monarch in summer looks much the same as a monarch in the fall. But the summer butterflies live only a few weeks while those of the fall store fat and fly south for the winter. Exactly what creates this difference and how the migrants know where to fly are only two of the mysteries still unanswered about these intriguing butterflies.*
JED MANWARING

LENDING A HELPING HAND

Monarch butterflies cannot survive very far into the next century as a widespread species across North America unless people take steps to change the things that endanger this awe-inspiring insect. This means putting a stop to logging and the destruction of wild plants in critical areas and creating more places where the butterflies can rest, feed, and breed in safety.

One option for saving the winter roosts in Mexico is for conservation groups or governments to buy the surrounding forest from the communities that own them. Logging companies are pressing to open these areas up, leaving little time for action to prevent further deforestation. Although logging brings the promise of employment and money to the thousands of of residents who live in the dry, hilly land around the reserves, it's a short-term gain with no future once the trees, and the butterflies, are gone.

A longer-term alternative for people in the area may be to develop a tourism industry. Few spectacles in nature can compare with the heart-stirring sight of millions of monarch butterflies blanketing the trees of their winter refuges. It is a sight that more and more people are eager to see for themselves, a fact that may benefit the butterflies if it brings money to the local economy. One of the reserve areas, el Rosario, is already open to the public and attracting a large number of visitors. It has a small information centre, and local villagers work as tour guides and interpreters as well as selling food and souvenirs. Tour operators literally truck tourists up the steep mountain slopes to see the treetop clusters of monarchs, but tourism is a development with a downside as well.

Since the butterflies stay in the roosts for only five months of the year, the reserves cannot provide year-round employment. Local organizations are working to overcome this problem, helping nearby residents diversify their economy and make a sustainable living from their environment in other ways, such as breeding fish or growing flowers or fruit. A more serious problem is that the small area and scant number of roosts in the region limits its capacity to meet tourist demand, and there is a big risk of visitors damaging the very thing they are coming to see. It's a common dilemma for nature tourism everywhere. During Mexican holidays in February 1996, el Rosario had as many as three thousand tourists on a given day, all using only one or two trails leading to the butterflies. Apart from the problems of soil erosion along the trail and of air pollution from vehicle exhaust, visitors could hardly avoid trampling butterflies underfoot!

North of the border in California, the much smaller and more widespread monarch winter roosts have been a source of tourism for years. Pacific Grove, on the coast between

San Francisco and Los Angeles, dubs itself Butterfly Town, U.S.A., and holds a festival each fall where local children dress in monarch butterfly costumes. A 1938 city ordinance makes it "unlawful for any person to molest or interfere with in any way the peaceful occupancy of the Monarch Butterflies on their annual visit to the City of Pacific Grove, and during the entire time they remain within the corporate limits of said City, in whatever spot they may choose to stop in." In nearby Santa Cruz, a monarch butterfly flag is hoisted the day the first monarchs arrive and is flown for six months until the last one leaves.

The tourism dollars generated by the monarchs, as well as local pride in the insects, help ensure their future place along the California coast, where the main remaining threat is the loss of roosting trees to developers. To save a privately owned monarch grove from development, voters in Pacific Grove in 1990 approved the purchase of the property by the city for a cost of $2 million.

There is little legal protection for either the monarch or its habitat in the United States or Canada, although three areas in southern Ontario were designated as monarch butterfly reserves in October 1995. The protected sites (Point Pelee, Long Point, and Prince Edward Point) are fingers of land projecting from the north shores of Lakes Erie and Ontario. Key points on the migration route, their reserve status was declared by Canada as part of an agreement with Mexico—the first time either government had taken formal international action to protect an insect.

International co-operation is essential for helping monarch butterflies, and should be made easier by the plans for environmental protection included in the terms of the North American Free Trade Agreement (NAFTA) signed by Canada, the United States, and Mexico. A trilateral Commission for Environmental Cooperation set up under this deal aims to promote more habitat protection, research, and monitoring of threatened species, but in the end it takes public education and public pressures to make such good intentions effective.

Probably one of the simplest and most effective ways for people to help the butterflies in the United States and Canada is habitat restoration. Many states and provinces in recent years have started growing flowering plants along roadside verges and median strips, replacing the costly and environmentally damaging practices of mowing and spraying used to maintain neat, grassy margins. In one move, this adds butterfly habitat, saves money, reduces pollution, and, as a bonus, is nicer to look at. The strips of wildflowers are easy to maintain, needing mowing only once a year, in late fall, to keep down the growth of woody plants. (Unfortunately, traffic itself will remain a major hazard for butterflies following the asphalt south, and many thousands of migrating monarchs are killed by vehicles along the way.)

Chrissy

Wildflowers and milkweeds can be encouraged to grow along powerline corridors and in provincial, state, and national parks, keeping these areas open by selective cutting or controlled burning of shrubs after the monarch breeding season is over, rather than by using herbicides. Chemical-free butterfly gardens with wild or cultivated flowers can be created almost anywhere, from an urban backyard to a school or museum.

Bringing butterflies and people close together is a powerful means of education, and children especially are fascinated to follow the career of a caterpillar, watching the life cycle unfold from egg to butterfly. In his excellent *Handbook for Butterfly Watchers*, Robert Michael Pyle shares his enthusiasm for these insects and explains how to observe, collect, rear, and photograph them. Pyle is founding president of the Xerces Society, an organization for promoting research and conservation of butterflies and other invertebrates, and he does much to help people of all ages understand the importance of butterflies in the environment.

There is usually no doubt among children that they want monarchs to remain a part of their world. It was a third-grade class in Decatur, Illinois, for example, that spurred their state to pass a law in 1974 making the monarch butterfly its "official insect." In Nebraska, students wrote letters to the state's Noxious Weed Advisory Committee to convince them of the importance of letting milkweed grow for monarch caterpillars. Without caring actions like these, and ceaseless vigilance, monarch butterflies could disappear very quickly.

It may seem alarmist to worry about the disappearance of an animal that still numbers in the tens of millions, but numbers alone give no immunity to extinction. The passenger pigeon was the commonest bird in North America during the last century, and one of the most numerous the world has seen. Its flocks numbered in the billions, and darkened the sky as they passed in flight over the forests and farms of the northeast. Shooting and trapping seemed unlikely to dent their vast numbers. As late as 1879, a billion passenger pigeons were captured in the state of Michigan. But the last nest of these unique birds was observed in 1894, and the last member of the species died in the Cincinnati Zoo in 1914. A combination of slaughter and destruction of the forests they depended on drove passenger pigeons from the face of the earth in little more than the span of a generation.

Monarch butterflies may yet share the sad fate of the passenger pigeon. Vast numbers are at immediate risk because they depend on a handful of sites in a small area of Mexico that is still being logged. They are at long-term risk because their breeding habitat throughout the United States and Canada is disappearing. The conservation of these butterflies clearly depends on the cooperation of three nations, united by NAFTA into one of the wealthiest trading areas of the world.

FACING PAGE *Every October, a parade of winged children marches in the streets of Pacific Grove, California, to welcome the return of monarch butterflies to their town for the winter. The tradition of the children's march dates back to the 1940s.* FRANS LANTING/MINDEN PICTURES

In the years to come, my chances of being transported once again by the sight of monarchs streaming along the shores of Lake Ontario depend on the will of governments in Mexico City, Washington, and Ottawa. They depend on the conditions of Mexican farmers on their forested slopes, and on the boardroom decisions of huge logging companies. They depend also on the individual actions of people throughout North America who choose to care about the fate of these beautiful and intriguing creatures.

FACING PAGE *More and more people are discovering the rewarding hobby of butterfly watching. Monarchs are among the easiest of species to identify.* DON RIEPE/PETER ARNOLD, INC.

Gardeners are butterflies' friends. Anyone can give these harmless and attractive creatures a helping hand by encouraging flowers to grow in their neighbourhood. THOMAS KITCHIN/FIRST LIGHT

FOR FURTHER READING

NATURAL HISTORY

Brower, Lincoln P. 1995. Understanding and misunderstanding the migration of the monarch butterfly (Nymphalidae) in North America: 1857–1995. *Journal of the Lepidopterists' Society* 49, no. 4: 304–385.

This scholarly review brings together the history of research on the monarch migration from the middle of the last century to today.

Herberman, Ethan. 1990. *The Great Butterfly Hunt.* New York: Simon and Schuster.

An account of the monarch's migration and the search for the winter roosts in Mexico. Written for young people, it includes many colourful photographs.

Lavies, Bianca. 1992. *Monarch Butterflies: Mysterious Travelers.* New York: Dutton Children's Books.

A book for younger readers, filled with beautiful photographs showing the monarch's life history and the roosting sites in Mexico.

Malcolm, Stephen B., and M. P. Zalucki, eds. 1993. *Biology and Conservation of the Monarch Butterfly.* Los Angeles: The Natural History Museum of Los Angeles County.

This collection of academic papers by leading butterfly researchers focuses on recent discoveries about monarch biology and looks at the issues surrounding their conservation.

Pyle, Robert Michael. 1992. *Handbook for Butterfly Watchers.* Boston: Houghton Mifflin.

A well-written and comprehensive introduction to watching, rearing, photographing, and understanding butterflies.

Urquhart, F. A. 1960. *The Monarch Butterfly.* Toronto: University of Toronto Press.

The original classic on the monarch butterfly. This book includes research data and describes the monarchs' habitat, life cycle, anatomy, behaviour, and migration.

Urquhart, Fred A. 1987. *The Monarch Butterfly: International Traveler.* Chicago: Nelson-Hall.

An updated version of the original book, revised for a general audience.

ADDITIONAL RESOURCES

Monarch Watch
http://monarch.bio.ukans.edu/

An Internet site of the University of Kansas. Postings on the site include a variety of news and basic information about monarchs, as well as details of tagging and other research programs.

North American Butterfly Association (NABA)
4 Delaware Road
Morristown, New Jersey, U.S.A. 07960
Telephone: (201) 285-0907

A group for anyone interested in butterflies. NABA produces a quarterly magazine, *American Butterflies,* with articles by experts on how to attract butterflies to gardens and how to identify them, and with the latest news from researchers. NABA's annual butterfly count (held across the country) helps monitor changes in populations.

The Xerces Society
4828 Southeast Hawthorne Boulevard
Portland, Oregon, U.S.A. 97215
Telephone: (503) 232-6639

Named for an extinct species of butterfly, the Xerces Society promotes education and conservation of all Lepidoptera and other invertebrates.

INDEX

A

Abdominal claspers, 31
Adult butterfly, 28, 29
Ambush bug, 83
Antennae, 11, 22, 27, 29
Apuleius, 28
Aridjis, Homero, 92
Asclepias (milkweed genus), 73.
  See also Milkweeds
Australia, 8

B

Baja California, Mexico, 57
Bates, Henry W., 77
Birds, as predators, 64, 74, 81
Black-backed oriole, 81
Black-eared mouse, 81, 83
Black-headed grosbeak, 81
Boyce, T. M., 64
Braconid fly, 83
Breeding grounds, 48 (map),
  57, 58, 68, 96, 103
Brower, A. V. Z., 64
Brower, Lincoln P., 53, 57, 61,
  68, 80, 92, 95
Brugger, Ken and Cathy, 53
Butterfly family, 8
Butterfly garden, 103

C

California, 8, 57, 100
Canada, 8, 44, 50, 58, 60, 67, 96, 101
Cardenolides, 73–74, 81, 83
Caribbean islands, 8
Cascade Range, 48 (map), 57
Caterpillar. *See* Larva
Central America, 8, 79
Cerro Pelón, 53
Chrysalis. *See* Pupa
Cleveland, Ohio, 39
Coleoptera, 7
Commission for Environmental
  Cooperation, 101
Courtship, 31–34. *See also* Mating
Cremaster, 18
Cupid, 28
Cuticle, 14–15, 18

D

Danaus gilippus. *See* Queen butterfly
Danaus plexippus, 2, 8
Darwin, Charles, 77
Decatur, Illinois, 103
Defence against predators, 15. *See also*
  Cardenolides *and* Warning colours
Deforestation. *See* Logging
Diet. *See* Food
Diseases, 84
DNA analysis, 61, 64

E

Egg, 10, 11–12, 14, 16, 48, 58
El Rosario, Mexico, 100
Epidermis, 5
Eucalyptus trees, 57
Europe, 8, 47

Evolution, 7, 47
Eyes, 16, 22, 29

F

Flight
  courtship, 31
  distance, 2, 36, 54, 57
  paths of, 58
  speed of, 29, 50
  See also Migration
Food
  of caterpillar, 11, 14–16, 73
  of adult, 44. *See also* Nectar

G

Generations, 16
Genetic analysis, 61, 63, 64
Great Lakes, 44, 47
Great Plains, 68
Growth, rate of, 14, 16
Gulf Coast, 61, 68

H

Habitat restoration, 101
Hairpencil gland, 31
Hearing, sense of, 16
Herbicides, 92, 96, 103
Hibernation, 40
Hormones, 8, 14–15, 16, 18
Human impact, 68, 84

I

Ice Age, 68, 92
Instar, 14, 15, 16
Internet, 60, 111
Intestine, 22

L

Lake Erie, 44, 101. *See also* Great Lakes
Lake Ontario, 40, 101. *See also* Great Lakes
Larva, 3, 8, 10, 11, 12, 14–16, 17, 83, 84, 96
Legal protection, 101
Legs, 11
Lepidoptera, 2, 7, 8
Life cycle, 8
  adult butterfly, 28, 29
  egg, 10, 11–12, 16, 58
  larva, 14–16
  pupa, 18–22
Life span, 34, 96
Limenitis archippus. See Viceroy butterfly
Logging, 68, 87, 88, 92, 95, 100, 103
Long Island, New York, 47
Long Point, Ontario, 44, 101

M

Male, difference from female, 31, 43
Mating, 11, 31–34, 36, 54, 83–84. *See also* Courtship
Megapopulation, 61
Metamorphosis, 8, 22, 28, 73
Mexico, 2, 3, 8, 39, 44, 46, 47, 48, 50, 53, 57, 58, 60, 67, 71, 81, 88, 96, 100, 101
Mexico City, 2, 53
Mice, as predators, 81, 83
Microclimate in winter roosts, 54, 57, 92
Migration, 2, 3, 36, 39–50, 54, 57, 58, 60
  map of fall routes, 48
Milkweed bug, 11
Milkweeds, 8, 10, 11, 15, 16, 47, 48, 58, 68, 73–74, 80, 83, 96, 103
  chemicals in, 15, 61, 73. *See also* Cardenolides

Mimicry, 77, 80
Monarch Watch, 60, 110
Monterey Peninsula, 57
Monterey pine, 57
Moths, 7
Moulting, 14–15, 18
Moulting fluid, 15, 16
Mourning cloak butterfly, 40
Müller, Fritz, 80

N

Natural selection, 77
Nebraska, 103
Nectar, 1, 7, 27, 58, 29, 44, 54
New Zealand, 8
North American Butterfly Association, 110
North American Free Trade Agreement (NAFTA), 101, 103

O

Ontario, 1, 2, 44, 61, 96, 101
Overwintering, 8, 50, 53–57, 92. *See also* Roosts
Oyamel forests, 53, 57, 92

P

Pacific Coast, 48, 57
Pacific Grove, California, 100–101, 103
Parasites, 83–84
Passenger pigeon, 103
Pheremones, 31, 43
Pleistocene Era, 47
Point Pelee, Ontario, 44, 101
Pollution, 95, 96, 101
Population
  distribution, 8, 48 (map), 61, 64, 68
  size, 53, 61, 64, 67, 68, 84, 87–88, 96, 103
Predators, 15, 16, 47, 64, 74, 81, 83, 84
Prince Edward Point, Ontario, 101
Proboscis, 7, 22, 27, 58
Psyche, 28, 35
Pupa, 16, 17, 18–22, 28
Pupa stalk, 18
Pyle, Robert Michael, 103

Q

Quebec, 96
Queen butterfly, 31, 77, 79, 80, 83

R

Red admiral butterfly, 40
Red-winged blackbird, 80
Reproduction. *See* Mating
Reserves, 92, 100
Ritland, David, 80
Rocky Mountains, 8, 48, 54, 57, 61, 64
Roosts, 39, 44, 47, 50, 53–57, 81, 92, 100

S

San Francisco, 57
Santa Cruz, 101
Senses
  hearing, 16
  sight, 16, 29
  smell, 11, 27, 29
  taste, 11, 27, 29
Sex differentiation, 31, 43
Shield bug. *See* Stink bug
Sierra Nevada, 48 (map), 57
Sight, sense of, 16, 29
Silk button, 18
Silk gland, 18

Smell, sense of, 11, 27, 29
South America, 8, 79
South Pacific islands, 8
Sperm sacs, 31
Spinneret, 18
Stink bug, 83, 84

T

Tachinid fly, 83
Tag, on wing, 50, 57, 60, 64
Taste, sense of, 11, 27, 29
Taylor, Orley R. (Chip), 60
Toronto, Ontario, 1, 2
Tourism, 100, 101

U

Ultraviolet light, 29
United States, 3, 8, 39, 47, 48 (map), 50, 53, 57, 58, 60, 79, 96, 100, 101, 103
Urquhart, Fred, 2, 16, 44, 50, 53, 61

V

Viceroy butterfly, 77, 80
Volunteers, 50, 60–61

W

Warning colours, 74, 77, 81
Weather, 5, 28, 29, 44, 47, 54, 84, 87, 88, 92, 96
West Indies, 61, 79
Wildflowers, 68, 69, 92, 96, 101, 103
Wings, 2, 8, 22, 28, 64. *See also* Tag
Winter survival, 39, 81, 87–88. *See also* Overwintering

X

Xerces Society, 103, 110